THE BULLY in your POCKET

YOUR #1 Playbook to DEFEAT ONLINE TROLLS

BY: LORENZO GOMEZ III

DEDICATION

This book is dedicated to my writing coach, mentor, and friend Barbara Boyd. Thank you for your guidance all these years. You have always been and will always be "the defender of the reader."

CONTENTS

Introduction 1

Section 1: Understanding Trolls & Online Bullying 11

    Chapter 1: Anatomy of a Troll 13

    Chapter 2: The Deadly Ds 21

    Chapter 3: The Last Word & Mob Rules 43

Section 2: My Fight with a Troll **53**

    Chapter 4: A Troll in Sheep's Clothing 55

    Chapter 5: One Bad Journalist 70

    Chapter 6: Dark Thoughts, Old Feelings 86

Section 3: The Playbook: How to Beat a Troll **97**

    Chapter 7: Pain For Me, Gold for You 99

    Chapter 8: The Toughest Row to Hoe 106

    Chapter 9: Boundaries Beat Bullies 135

Conclusion **146**

Acknowledgements **156**

About the Author **158**

Notes **160**

INTRODUCTION

A Haunting Question

"The wise man doesn't give the right answers, he poses the right questions."

– Claude Levi-Strauss

"What is the biggest difference between going to school now and when you were going to school?"

I reached into my pocket and pulled out my iPhone.

"This is the biggest difference," I said. I raised it higher for the entire group to see it.

"When I was in school, if someone was bullying me, all I had to do was get home. If I could get home, I knew I was safe. My father & three brothers are the toughest guys I know. If I could just get home, nothing would matter because no one could hurt me."

"You all have something much more difficult to deal with. All I had to do was get to my house, go to my room, close the door, and be safe. But this phone has the ability to bring all that bad stuff into your bedroom with you."

"This phone can be the greatest tool for learning, socializing, and communication, or it can be the *bully in your pocket*. You can go into your bedroom, into the most intimate place where you are supposed to feel safe, turn it on, and there waiting for you are 200 classmates stabbing you with words and trying to murder your self-esteem. And all of this can happen without your parents or any other adult ever knowing about it."

"This is the grand challenge of your generation."

THE DAVID ROBINSON FELLOWS

In 2019, I was invited to speak to a group of high school students in the prestigious David Robinson fellowship at IDEA Carver Academy in San Antonio, Texas. The Carver Academy was founded in 2001 by legendary Spurs Hall of Famer David Robinson, aka The Admiral. San Antonio is majority-minority with a mostly Hispanic population. The Carver Academy was established in its poorest black community located on the east side of downtown. Idea Public Schools is a charter school network which started in South Texas and then expanded north.

The David Robinson fellowship is special and an exceptional fellowship program. Ten Carver students are admitted into the program. Over their four years, they will receive individual mentoring and financial support from "The Admiral," David Robinson himself, the legendary U.S. Naval Academy graduate.

The school specifically asked me to speak to the fellows during their mental health session. I had just published my second book, *Tafolla Toro: Three Years of Fear,* which is a mental health book about my middle school years. So, when the email came asking me to speak at this session, I was honored to accept. Also, when you're asked to do anything on behalf of The Admiral, you just say yes.

Having done public speaking for a couple of years, I can say confidently that K-12 students are the most challenging audiences to speak to. Adult audiences at least pretend to pay attention and will give you a small giggle at your bad jokes when they don't land. Not so with young people. If your joke sucks, expect it to land with a thud. Generally speaking, the girls are always more attentive, and the boys have their arms crossed, pretending to be too cool for school. This is normally how it goes with those speeches, but not this crowd. The Robinson Fellows lived up to their reputation as the future leaders of this city. They were locked in with laser focus from the moment I started speaking.

I gave my speech and then moved to Q&A, which is when students really come alive. Young people have asked me some of the most thoughtful and difficult questions if they feel comfortable enough to ask. In this case, the Robinson Fellows immediately started firing questions at me, and the game was on. Like verbal Jiu Jitsu, I was prepared for every question and felt my confidence growing as they lobbed new ones at me. Then something happened. A girl in the back of the class asked me a question that stopped me dead in my tracks. It was a simple question.

She asked, "What is the biggest difference between going to school now and when you were going to school?"

That, my friends, is the million-dollar question.

As my brain registered the enormity of her question, I must have stood there silent for about ten seconds. But for me, it felt like an hour. I knew the answer instantly but did not know how to express it. My mind was processing so much data it couldn't decide which thought to say first. I could feel my phone growing heavier in my pocket, almost like it was crying out to me, telling me it wanted to jump out and answer her question for me.

As I stood there, I looked at her, but I wasn't really there. In my mind, I was watching a movie that had a thousand tiny little scenes flashing by me at a million miles a second. They were scenes of angry comments and shared posts written in all caps. There were scenes of obituaries, heartbreaking head-lines, and GoFundMe campaigns for victims. I saw the faces of people I knew and the profile pics of complete strangers. And I saw the endless scrolling of angry and hateful content page after page.

It was almost as if my subconscious had been waiting for someone to ask the question, as if all these scenes have been silently stockpiled for this very moment. It wasn't until that moment I realized we all had a potential bully in our pockets.

Then I reached into my pocket and pulled out my iPhone.

WHY THIS BOOK NOW?

There are two reasons that I decided to embark on this adventure to bring civility and respect back to the internet and help people protect themselves from the army of trolls that now plague the online world.

The first reason is that this young woman's simple question ignited something in my brain that day. When she asked me that question, a seed was planted in my mind that has slowly grown over time. With each passing day, it has taken up more and more of my thought-life. It has gnawed at my brain, driving me so crazy I couldn't take it anymore.

It's a question so thoughtful that it deserves an entire category of TED talks. There should be an entire field of study, with millions spent on research, and college courses to answer her simple question. I'm sure after I publish this book, someone will tell me these things exist, and if so, that is great.

Mostly, I realize my real answer to her question, the unabridged version, is this book. Let this book be my contribution to the tools, strategies, and tactics on how to take back the online world and make it a safe place once again for us to explore without fear.

The second reason is rooted in my birth year. I was born in 1980, which is normally pretty insignificant except when you consider the topic of this book. Being born in 1980 makes me part of Generation X. So, what does that have to do with trolling, online bullying, and the internet? Well, everything.

Being born in 1980 means I am part of the very special generation that remembers the old and the new. I remember every house having a landline phone. I grew up with no computer in my house. I did not get a cell phone until I was 18 years old and I could pay for it. Back then, I had to go through a credit check before they would give me a cell phone. It's also because cell phones were barely becoming affordable enough for the mass market. I was in high school when the internet really started taking off. I remember classrooms having computers, and I remember using them but not to get online.

It wasn't until my senior year in high school that I discovered a company called Napster allowed you to download music and not pay for it. All the record companies were losing their minds. This was the very first industry I saw disrupted, and realized this thing called the internet was no joke.

Being born in 1980 also meant if I wanted to ask a girl out as a kid, I wrote a note that said, "Do you want to go around? Circle Yes or No," walked up and handed it to her. I didn't get a Facebook account until 2004, when I was 24 years old and working in the UK at Rackspace, a tech company serving online businesses. By then, I was a working professional who was translating old-school social skills into the online world, like asking a girl on a date via direct message (DMs) in addition to in person (I'm still old school like that).

Why does this matter? It matters because as a Gen Xer, I am of the generation who had to translate the old ways into the new world. I grew up in the very rare time when my peers and I had to take everything we did in the real world and transfer those skills, functions, and behaviors into the new online world. And we had to do it in real time, as they were evolving.

My entire life was training me for this moment. As my friends and I transferred all these real-world situations to the online world, we discovered what worked and what didn't. What needed to be adapted and changed. But nothing I did as a Gen Xer could have trained me for the birth of trolls and online bullying. That is, not until I started working for an internet company called Rackspace.

THE DISCLAIMER

From this point on, I am going to be using a combination of publicly accessible stories as well as incidents I have either experienced personally or witnessed. For all people involved in my personal stories, I have changed certain names and details out of respect to the individuals involved. The last thing I want is to write a book about online bullying and have it lead to more people being bullied. My desire is not to call anyone out but to use these stories as examples we can all learn from.

Each of these stories has contributed to the tools, tactics, and strategies I have developed to give you. They are designed to help you defend yourself but not to cause harm to anyone. Because of this, I have fictionalized many parts of my personal stories where I feel appropriate.

My second disclaimer is that this book does not bash the internet or social media companies. I consider the advent of the internet, email, and social media to be inventions so world-changing, I rank them right up there with the creation of the light bulb, electricity, democracy, and my personal favorite, because I live in Texas, air conditioning.

Having said that, every good thing can be turned into a bad thing, and these online tools are no different. In order for us to get solutions, I am going to point out some blind spots they have and tell you some hard truths, not to throw stones, but so we can grow and get better.

Also, since the time of this writing, Twitter has undergone a name change to X. But for the purposes of this book, I will refer to it by its previous name of Twitter, which is what most people are familiar with.

Lastly, this book contains stories of how I was bullied personally and some of the ways I responded, which I am not proud of. I share them with you in full transparency so you can see how high the stakes are, so you can learn from my mistakes and not repeat them.

The strategies and tactics I will lay out for you in this book absolutely work. How do I know? I know because I had to learn them the hard way. It is my desire to save anyone reading this book the pain I had to go through to acquire this knowledge. I have used them in the real world, and I have advised others on using them. They work every time if you can maintain the discipline to stick to them. On one hand, they are very straightforward and simple, and on the other hand, they are hard as hell.

My desire is to pass them along to you so together, we can start a movement and create a new type of online community, one we will all be proud to be a part of.

SECTION I

Understanding Trolls & Online Bullying

CHAPTER 1

Anatomy of a Troll

"A fool's lips walk into a fight, and his mouth invites a beating."

- Proverbs 18:6

Let's start with the first question. *What is a troll?* As my branding mentor once told me, the simple message always wins. In this case, defining a troll is no different. Urban Dictionary has a pretty good definition, which is:

"Someone who deliberately pisses people off online to get a reaction."

The Lorenzo definition is even simpler.

"A troll is an online bully."

It is plain and simple. Throughout this book, you will see me use the words "troll" and "bully" interchangeably.

They can be the customer who flames a restaurant in their Yelp review, making it personal and not constructive. The celebrity who uses Twitter to cancel a person or company.

Even journalists and news outlets can be trolls. For most of us reading this book, it will show up as a friend, acquaintance, or family member who loves to comment on your posts just to get a reaction from you and everyone else who is watching.

Then there is the person who is almost a troll. They are not one yet, but they are in the middle of a transformation. I see this often with new users who have avoided the internet and finally join by creating an account on one of the platforms, like Facebook. As the normal human desire to be seen and heard gets stronger, they also get more aggressive with their post. They don't really want conflict, but posting something controversial seems the only way to get people to interact with them.

These people are going through a transformation online. They are like caterpillars; their metamorphosis will not turn them into a beautiful butterfly but rather a fire-breathing troll. I add them to this list because there is still hope for them. And I hope this book will encourage them to reverse their course and take a different path in their online behavior.

The final type of troll I will discuss is the one you will see the most in this book — the person or persons who design and launch a coordinated campaign against a person, group, or organization.

These are the people who peddle hate in bulk. They are the terrorists of the internet and are as sinister as a drug cartel or the Mafia.

This form of trolling is the most toxic because it is carried out with a predetermined goal of destroying the person or thing on the other end. They offer no quarter to their victims. They seek complete destruction whether it is emotional, financial, reputational, or physical.

In a time before the internet, we would have called these people bullies. However, the old label of bully is not strong enough for this new behavior. The bullies of old would harass you, make fun of you, and even, in extreme cases, maybe beat you up. But in the old-world, black eyes heal. In the new world, many of these wounds do not go away. In this new world, marriages are destroyed, careers ruined, and even suicide or death. On top of that, there is now a digital record of the entire ordeal which will live forever in some computer database on the internet.

But I don't need to convince anyone reading this book that we have a problem. The hard part is knowing how we defend ourselves against this new threat. I am here to confidently tell you that trolling is something you can overcome and defend yourself against.

Step one is understanding the mind of a troll and accepting certain parts of their personality that you cannot and will not change. Once you understand how they think, you will be ready to face them with more realistic expectations and, more importantly, the upper hand.

KEYBOARD KILLERS

Keyboard crusaders, keyboard cowboys, and keyboard warriors are the most popular names they have been called. You may think my rebranding them to Keyboard Killers is a tad dramatic, but I do this with great intention. The original trolls of yesteryear may have been a more passive version of the ones today, but the trolls who roam the streets of the internet today are not.

The reason I have renamed them is to put forth a simple and important idea. In today's online world, the modern-day troll has one desire — to hurt you. And over the last ten years, the world has watched as they have stacked up a body count so massive it should make everyone concerned.

Behind the keyboard, everyone feels like they are 6 foot 3, 250 lbs, bullet proof, fighting for the UFC. They are the people who post stuff so mean and so vicious you sit back and say, "Wow, I can't believe they just said that." But to truly understand them, we must understand their history. And like everything else in life, I had to learn about them the hard way, by trial and error.

When I was 20 years old, I was hired by a managed hosting company called Rackspace. What is managed hosting, you may ask? It took me 20 years to explain to my friends and family, but let me say that it's sort of the plumbing of the internet. We would host computers (called servers), and businesses would put their websites on them.

I first started working at Rackspace in August of 2001. Back then, I didn't really understand the internet other than the fact it was a big deal, it was used to send email, and that Napster had punched all of the record companies in the face for making us pay $20 for a CD that cost $2 to make and had only one song that we wanted. So, I was down with this kind of technology that lets us stick it to the man.

Back when Rackspace was first founded in 1998, its primary service offering for hosting was Linux servers. For those of you who are not tech geeks, just know back then, most websites sat on a computer somewhere with one of two operating systems: Microsoft Windows Server or Linux Redhat. Linux is an open-source operating system, which means it is free for anyone to download, and because it was free, it was the most popular one on the internet.

But even if you don't understand any of that, all you need to know is this. Even though Linux was free, it was very difficult to learn. That meant the Linux customers who paid us to host their websites were the geeks of all geeks. They were power geeks, and I would say they laid the foundation for the modern-day internet. Their websites, applications and code are what expanded the universe of the internet. So, in short, they were smart, prickly, and they did not suffer fools. Especially little 20-year-old account managers who were still developing their technology skills like me.

I was dealing with a group of people who were used to everyone around them not understanding what they did for a living. And because of this, they felt extra safe to do and say whatever they wanted. For me, this was the birthplace of the keyboard crusaders. I didn't know it at the time, but I was interacting with a new species of internet user which was evolving and adapting as the internet expanded.

I would also submit to you that it was from within this group of power geeks that the internet ancestors of trolls were born.

THE CHRISTIAN DATING WEBSITE

My first direct run-in with a keyboard killer came from one of my customers at Rackspace. The website we were hosting was The Christian Dating Website, and the primary contact I interacted with was a guy named Kevin. I came in one Monday morning, and my entire support team was abuzz over something that had happened overnight.

I logged into our system and opened up my email. As my email loaded, I realized something crazy had happened because I had dozens of automated emails from our customer support and data center teams about one of my customers. I looked up their account to see which customer, and low and behold, it was The Christian Dating Website. I read through all the emails, but all I could determine was that their website had stopped working sometime during the night. Our team had got it back up several hours later.

It didn't explain why there was such a stir, though. Websites went down all the time. Although it's not a good thing, it had never gotten everyone riled up like this. I decided to get to the bottom of it. I walked over to my mentor, Omar's, cubicle. Omar was wise, smart, and always got to the office really early so he could chat with the overnight engineers. I knew Omar would give me the scoop.

"Omar, what's going on with Christian Dating?"

"You didn't hear?" Omar said, laughing.

"No, what happened?"

"So apparently, the Christian Dating website went down around midnight, and no one caught it for a couple of hours. Then their primary contact, Kevin, called in screaming at everyone."

He said, *"I'm losing thousands of dollars every minute, and if you don't get my site up asap, I'm going to come down there, and I'm going to KILL YOU!!"*

"What?! No way. He did not say he was going to kill them. "

"Yeah, dude, he literally said he was going to kill the 3rd shift guys if they didn't get his site back up."

I laughed. I couldn't believe how absurd this story was.

"I guess he wasn't wearing his 'What Would Jesus Do' bracelet."

I walked back to my desk. My stomach turned as I realized I needed to call Kevin and follow up with him to make sure everything was back up and running smoothly. But how does a 20-year-old approach someone who just threatened to murder his teammates. I was terrified to call him, but I did what I was trained to do. I picked up the phone and called.

"Hello?"

"Hello, is this Kevin from The Christian Dating Website?"

"Yes, it is."

"Hey Kevin, this is Lorenzo Gomez from Rackspace. I'm calling to followup with you on your recent issue. I understand your site went down last night, and you had to call in to get it back up. Is that correct?"

"Yes."

"Well, per our service agreement, I am going to issue you a refund for the time your site was down, ok?"

"Sure, that's fine."

"Is there anything else I need to know or can help with?"

"No."

"Ok then, I apologize for the trouble this has caused you. If you need anything else, please don't hesitate to call me."

"I will. Goodbye."

I hung up the phone more confused than before. None of it made any sense to me. How can a guy who just threatened to kill an entire group of people a couple of hours ago be as calm as a Hindu cow the next?

As it turns out, the answer lies inside three little illnesses that have sprung up in the online world. For my angry customer, he had lost his empathy for a short period of time, and he thought there would be no consequences to him threatening to kill my teammates.

These are just two deadly illnesses I wish to explore next.

CHAPTER 2

The Deadly Ds

"Likewise, the tongue is a small part of the body, but it makes great boasts. Consider how a great forest is set on fire by a small spark. The tongue also is a fire, a world of evil among the parts of the body."

James 3:5-6

With a high degree of confidence, it can be said that the internet has given the world many amazing innovations. It is also the most powerful tool of the modern world. It offers many different ways to connect and interact with people than we had previously. When I lived abroad, I met so many amazing people I'm still friends with to this very day. We are able to communicate even though they live on different continents and in different times zones. If I want to text them, I just use WhatsApp. If I am traveling to their country and I want to meetup for coffee while I'm there, I DM them on Facebook or Instagram. That would not be possible without the internet.

However, human beings seem to have a gift for taking really amazing things and finding a way to corrupt them. I am not being a pessimist; this is just a historical fact. The internet is no different. As it grew and people realized its endless potential, it was hailed as the tool that could fix so many grand challenges in our society. While this is true, we didn't expect a deadly cancer to grow.

But before it developed into full-blown cancer, it started out as three little illnesses. They are what I call "The Deadly "Ds." They are deception, denial, and distortion that trick your brain into thinking a certain way. The third D, distortion, actually recalibrates something in your mind and alters your view of humanity. They are:

The Deception of Anonymity
The Denial of Consequences
The Distortion of Empathy

These are the three sicknesses that have infected the online world. And just like the zombie apocalypse, they are highly contagious. However, unlike the zombie apocalypse, they are completely preventable and curable. In this chapter, I will walk you through them and what it sounds like when we are engaging in each one. Once we know how they work, we can adjust our behavior to remove their destructive power.

But before I walk you through each of them, I need to tell you a tragic story that took place in my beloved hometown of San Antonio. The reason I picked this story is because it has to do with online bullying, and it is a case study of how the Deadly Ds show up in almost every situation like it.

It is the story of a young man named David.

DAVID'S LEGACY

David Molak[1] was the kind of teen any parent would be proud of. David was an Eagle Scout, an avid San Antonio Spurs fan, and was even named "Athlete of the Month" at his local gym[2]. He is also the tragic victim of online bullying, whose story I hope can help shed light on how these situations unfold and how the future will never be the same.

In 2015, David Molak started getting bullied by classmates at his local high school in San Antonio, Texas. In October 2015, a classmate of the 16-year-old sophomore posted a picture with David's girlfriend on social media with a caption saying he was going to steal her away from David. They tagged David in the post and soon hundreds of comments followed.

One student said, *"Molak's an ape."* One student followed with, *"The monkey looking human gets his woman stolen."*[3]

Another teen posted *"Put um inna coffin,"* and another said, *"Put em 6 feet under."'*

David's mother Maurine said of the event, *"He (David) said he could never go back to school. He just felt like everyone hated him."*[4]

Distraught by the harassment, David's parents pulled him out of school and transferred him to another high school in town. But it didn't stop the harassment from following him wherever he was.

One Sunday night, David and his brother Cliff were hanging out together and David got really quiet after reading a series of text messages he had received suddenly.

"*A set of 6-10 unknown numbers added David Molak to a group text in which he began receiving comments berating him on his physical appearance,*" Cliff Molak recalled."[5]

In the end, it all proved too much for David. He had previously attempted suicide twice and on January 3, 2016, he took his own life.

This story is a tragedy and every family's worst nightmare. What is so astonishing about it, though, is that from the ashes of this horrible event, the Molak family decided to do something I believe will end up changing the landscape of online bullying across this country.

After David's death, the Molak family launched the David's Legacy Foundation, which is a nonprofit whose mission is to "*eliminate cyber and other bullying, of children and teens, through education, legislation, and legal action.*"[6] And in 2017, the Texas 85th legislative session passed SB 179 known as "David's Law," thanks to the hard work and advocacy of the Foundation.

I will go into more detail about this law in the next few sections but it basically empowers schools to investigate and take action against bullying. It also gives parents more legal tools to take action when the schools cannot or fail to do so. In short, it is a game changer and the start of what I feel will be a wave of similar laws to come all across the US.

The reason I share this sad and tragic story with you is because I would like to break it down and show you how online bullying works from a behavioral and psychological standpoint.

What happened in this story can happen to anyone at any time. It is my hope that it doesn't and my intention is now to show you what is happening inside the human mind when situations like this occur.

I would like to break down this story and a few others with the help of some psychological experts to show you what has become the internet's most dangerous online illnesses.

Illnesses I call *The Deadly Ds*.

THE DECEPTION OF ANONYMITY

The Lie: *"I am one out of a billion users online. It is easy to hide. No one will know who I am and no one can find me."*

In 2008, PayPal, the largest online payment system in the world, acquired a small, fifty-person Israeli company called Fraud Sciences for an estimated $169 million.

In Dan Senor and Saul Singer's book, *Start-up Nation: The Story of Israel's Economic Miracle*,[7] they explain how, before they purchased the company, PayPal's president, Scott Thompson, gave the Fraud Sciences team one hundred thousand PayPal transactions to analyze. Thompson wanted to compare Fraud Science's ability to spot fraudulent transactions against his own team at PayPal. What Thompson thought would take them months to analyze, Fraud Sciences did in three days. Not only did they do it faster than expected, but they were also more accurate than PayPal's team.

"The difference was particularly pronounced in the transactions that had given PayPal the most trouble – on these, Fraud Sciences had performed 17 percent better."[8]

How did they accomplish such an impressive feat? When asked how their methodology worked, in the book, the co-founder of Fraud Sciences, Shvat Shaked, said,

*"Good people leave traces of themselves on the Internet – **digital footprints** – because they have nothing to hide....Bad people don't, because they try to hide themselves. All we do is look for the footprints."*[9]

Shvat explains that he learned this skillset while in the army, assigned to a unit whose job was to hunt down terrorists by tracking their online activities. He said,

"Terrorists move money through the web with fictitious identities."[10]

So, if this concept has never been explained to you, let me break it down. Whether you know it or not, and whether you like it or not, you and I and everyone else online has a **digital footprint**. Most of us have never thought about this concept because we are not trying to hide our digital footprint because we are not doing anything wrong. In the early days of the internet, it was easy to hide, and no one could find you. These days, you can still hide, but it's getting harder.

However, in the not-so-distant future, there will be nowhere to hide.

The problem with this concept, however, is that most people don't realize this is where the internet is heading. So many people grew up using the internet in a way that made them feel like they could hide in the crowd.

Twitter historically has been a great example of this because you can literally talk at or "tweet" at anyone who is on the platform. You want to tell your favorite basketball player how crappy his free throws are, go ahead. You want to tweet at the President of the United States and tell him that he sucks or is awesome. Totally doable. Twitter's former CEO Jack Dorsey has said it is akin to a public square where anyone can speak their mind. And while I have mixed feelings about this comparison, it proves anonymity's deception.

If the internet and platforms like Twitter are the public square, thinking you are anonymous is easy. Anyone from anywhere and any background can walk up to you, interrupt your conversation, yell, spit in your face, and walk away without you knowing who they are. And those who behave badly feel emboldened to spew their venom and then casually stroll back into the crowd to hide from the fallout.

Now, let's get back to David's Legacy.

How did these bullies get away with such an aggressive campaign of harassment? Well, whether they knew it or not, they were using the deception of anonymity. Three days after David Molak ended his life, his brother Cliff wrote a moving post on Facebook that got shared 19k times. One of the things he wrote was about people's ability to hide online. He said,

"In today's age, bullies don't push you into lockers, they don't tell their victims to meet them behind the school's dumpster after class, they cower behind user names and fake profiles from miles away constantly berating and abusing good, innocent people. The recent advances in social media have given our generation a freedom of which has never been seen before."[11]

The deception of anonymity is the gateway drug to the second Deadly D. Once you think no one can find you, it emboldens you to do things you shouldn't because you think you can get away with them. In short, you come to believe that your actions have no consequences.

But as we just learned from our friends at Fraud Sciences, you cannot hide forever. The direction the internet is going is one where you will be found if you do something wrong, just like in the physical world.

So, what is the antidote for The Deception of Anonymity?

Act like anyone and everyone can see you. Write as if you will have to read that same post in a courtroom. You don't have to live a paranoid existence, but you need to act like the person you love and respect the most is watching you, not just the authorities. If your mother or grandma raised you, you need to act like they are watching, because you never know - Someday, they just might see what you did and posted.

Only, by then, it will be too late.

THE DENIAL OF CONSEQUENCES

The Lie: *"My actions in the online world have no real consequences in the physical world."*

On July 15, 2020, a major online cyber-attack was launched targeting 130 Twitter accounts of very high-profile people.

"The security breach saw accounts including those of Barack Obama, Elon Musk, Kanye West, and Bill Gates tweet a Bitcoin scam to millions of followers," according to an article by BBC.[12]

Basically, hackers took control of these famous people's Twitter accounts and started posting a scam asking people to send them money in return for more money. Below is the post taken from Joe Biden's Twitter account that day.

"I am giving back to the community. All Bitcoin sent to the address below will be sent back doubled! If you send $1,000, I will send back $2000. Only doing this for 30 minutes."

The hackers were basing their strategy on a simple numbers game. The people they targeted had so many followers they only needed a small fraction to take the bait. You may have read the post and correctly thought it is an obvious scam, but when you are dealing with people who have tens of millions of followers, you only need to fool a handful.

Just between Barack Obama and Elon Musk, they have over 250 million followers as of the time of this writing.

Even though the scam was fast and obvious to most, the hackers still received more than $100,000 of people's money.

The person or persons behind this attack were not only professionals, but I would also bet they firmly believe in the denial of consequences. They believed in it so much that the group was famous for hacking into people's accounts, finding compromising photos, and threatening to release them if they didn't send them money or post messages on their behalf.

Then, in July 2021, police arrested Joseph James O'Connor, aka Plugwalk Joe, from Liverpool, in connection to the Twitter hacks. He was facing 14 charges, including internet fraud, money laundering, extortion, and stalking. I want to reiterate that he was only accused and, in the US, you are innocent until proven guilty.

And this, my friends, is where I want to explain how the denial of consequences will slowly become a thing of the past.

Now, let's look a little closer at Plugwalk Joe. For those of you who don't know, instead of using his real name, Joseph Connor was using what is called a "handle." This is a nickname chosen by a user to identify themselves online. They are also referred to as usernames and they're used to provide anonymity and privacy online.

He was also accused of swatting a U.S. law enforcement officer. For those of you who don't know, "swatting" is the act of illegally calling law enforcement to report a false emergency at a target location with the intent to create a tactical response by law enforcement. [13]

You read that right. You piss off some kid playing Call of Duty, they find your location, call your local police department, and say you are about to do a mass shooting. Then boom, ten minutes later, the SWAT team shows up. It's crazy to imagine, but this actually happens.

I list all this stuff because not only is it pretty extreme, but it's also so obviously bad and against the law. So why would a person or group of people do these kinds of things? The answer is because they think they can get away with it. That or they are psychopaths, but that is a story for another book.

In his defiance, Plugwalk Joe told *The New York Times* before his detention, "*I don't care – they can come arrest me. I would laugh at them. I haven't done anything.*"

Now, I'm no legal expert, but I can come up with ten ways off the top of my head to say, "*I'm innocent,*" better than that. Just saying.

Here is the point I'm making for anyone out there who thinks you can get away with bigger and bigger "goofs or pranks" online. I am here to tell you the good old days are about to be long gone. Now, authorities and governments take online behavior very seriously.

In the case of Plugwalk Joe, just consider this: He is a British citizen who was arrested in Spain. A Spanish court approved his extradition, and he was sent to the US for trial. Three different countries were involved in something and someone who is allegedly involved in breaking the law online.

Then, in May 2023, Plugwalk Joe pled guilty in New York "*to a role in one of the biggest hacks in social media history,*" according to the BBC[14]. The hacking charges carry a total maximum sentence of over 70 years.

The US Assistant Attorney General, Kenneth Polite Jr., said,

"Like many criminal actors, O'Connor tried to stay anonymous by using a computer to hide behind stealth accounts and aliases from outside of the United States.

But this plea shows that our investigators and prosecutors will identify, locate, and bring to justice such criminals to ensure they face the consequences for their crimes."[15]

The consequences of online behavior now have global consequences.

Now, let's get back to David's Legacy.

I just showed you how the denial of consequences is quickly becoming the reality of consequences. Not only that, they are consequences on a global level. Let's talk now about consequences at the local level.

After David Molak's tragic death, his family successfully lobbied the Texas Legislature to pass a new law known as David's Law.

This law is very interesting because it gives both schools and parents more tools to help them combat online bullying. I will not go through the law extensively; for those interested, there is a great one-page summary on the David's Legacy website.[16] However, I want to point out a couple of items the law accomplishes.

For schools, it does the following:

- Expands the definition of "bullying" in the Education Code to specifically include cyberbullying.

- The Law applies to bullying occurring on school property or at school-sponsored activities, on school buses, and off-campus if it substantially disrupts the orderly operation of a school or interferes with a student's educational opportunities.

- Authorizes schools to remove students engaging in bullying activities from class, place them in disciplinary alternative education programs, or expel them for certain types of bullying activities.

- School personnel are authorized to report certain bullying activities that rise to the level of being a crime to law enforcement officials and strong protections from civil or criminal liabilities and disciplinary action are given to schools and school personnel who report criminal bullying to law enforcement officials under this law.

For students and parents, it does the following:

- Cyberbullying victims who are minors or their parents can seek injunctive relief against the cyberbully.

- Victims can ask the court to issue an injunction against not only the cyberbully, but also against the cyberbully's parents.

- It's also worth noting that cyberbullying is a Class A misdemeanor.

So, what does this all mean? In short, this is a game changer for both schools and parents. In the past, and I'm sure today, to a certain degree, most schools do not want to get involved. The last thing they want is the liability of some person suing them.

Truth be told, they will never be free of that reality. But this law gives them new protections from it.

However, the biggest change I see are the tools it gives parents and students. The fact you can get a local Judge to issue an injunction (court order) now changes the entire dynamic of these situations.

So, parents, if your child is getting bullied in a Texas public school, know this. You can get a court order requiring the bully to stop. It also requires the parents of the bully to take action to prevent their child from cyberbullying. If they don't stop or if they violate the court order, they will face legal consequences, such as fines or even jail time.

I had coffee with a representative from the David's Legacy Foundation after the law was passed. She told me a lot of times, parents won't really do anything if their child is bullying someone. But as soon as they receive a court order, everything changes, and the bully stops almost immediately in most instances.

Now, keep in mind this law I'm describing is in Texas and only applies to public schools in Texas. If you want to find out your state's policies, you can visit cyberbullying.org[17] for more information.

Finally, why didn't Kevin from The Christian Dating Website scream at me when I called him back later? It was because my phone call was the closest thing to a consequence he was going to get. And he was praying to God I wouldn't bring it up. Keyboard killers usually bark loudly until you call them on the phone or see them in person. Then they backflip like an Olympic gymnast.

But here is the point. The era of no consequences is coming to an end, and it's coming fast. So, as this new reality sinks in, let us now draw our attention to why people feel they can act like this in the first place.

Let us dive into the last of the Deadly Ds.

THE DISTORTION OF EMPATHY

The Lie: *"I see people online first as avatars, then as objects. And because they are objects, I no longer have to treat them as human beings."*

Note: An online avatar is defined as, *"An icon or figure representing a particular person in a video game, internet forums, etc."*

Let's return once more to The Christian Dating Website, my murderous Rackspace customer for a second. How could someone who is clearly very smart and technically savvy lose his sanity for a moment and threaten to kill people? The answer is that he had lost his empathy for a short period of time. And during that window, he only saw Rackspace employees as objects standing between him and his income. When it happened, he disregarded the humanity of others and was willing to treat them as objects to be removed, violently if need be.

Sadly, my story is not new to online users. We have all witnessed people say things that no sane person would say to another civilized human being. *"I hope you get cancer." "You should kill yourself."*

How does someone get up in the morning, kiss their spouse or children, log online, and stab friends and strangers with their words?

I was helped to answer this question by the book *The Science of Evil*, by Simon Baron–Cohen. He was troubled by the seemly endless cruelty human beings are capable of and set out to find a scientific explanation. What I found most interesting is that the most famous example people use is how the Nazis treated and tortured the Jews during the Holocaust, but as Cohen shows, it is a great mistake to think this kind of human evil was exclusive to the Nazis. The pages of human history are full of plenty of examples from all over the world.

One of the most powerful points he makes in the book is that in order to understand human cruelty, you must first see the other person as an object, and then you undergo what he calls "Empathy Erosion."

He says, *"Empathy erosion can arise because of corrosive emotions, such as bitter resentment, or desire for revenge, or blind hatred, or a desire to protect."*[18]

He continues by saying, *"The insight that empathy erosion arises from people turning into objects goes back at least to Martin Buber, an Austrian philosopher who resigned his professorship at the University of Frankfurt in 1933 when Adolf Hitler came to power."*[19]

"People turning into objects." This is the key insight.

After turning people into objects and removing empathy from your social interactions, you have unlocked the door to deny people their humanity.

Although I agree with Cohen on empathy erosion, what I am talking about is something completely different. In empathy erosion, your empathy goes away. I called it a distortion because I don't think people lose empathy on the internet. Something happens to your brain that temporarily turns you into a different version of yourself. I also think that while you are engaged online, something turns off the part of your brain responsible for seeking and expressing empathy. In empathy distortion, your empathy stops and then comes back.

This is why you can see someone you know acting and saying really horrible things online, but when you see them in person, they are a completely normal and lovely human being.

It is very similar to what morphine does to relieve pain. Morphine is an opioid that blocks pain receptors in the brain and essentially tricks your body into believing there is no pain. And that is what this is. The distortion of empathy is the morphine of the internet. It reaches inside your brain and crosses all the wires, even turning them off for a period of time. Once it does, you have become a distorted, mutated version of yourself, breathing fire on everything and everyone.

This happens online very subtly. First, when we create an online profile, we are required to upload a profile picture. This profile picture then becomes our avatar, or the virtual representation of ourselves. Some people don't bother to put a picture but instead an animated character or cartoon version of themselves. This actually makes it easier for other users to see us as objects and not as human beings. Now that you are reduced to an object, I feel empowered to attack and even murder your character, reputation, and your self-esteem with my words.

Always remember the road to human cruelty starts first in the mind, then is given birth by words, and finally manifests into actions.

For instance, let us briefly examine one of the most famous examples of cruelty starting out in the mind and then crossing the chasm into the real world. There is no clear consensus among historians on where Adolf Hitler's original hatred for Jewish people came from. But we do know it was in his mind many years before he was in politics. It was first born in his mind and grew there for years, like a virus.

Then, in 1925, he published *Mein Kampf*, where his thoughts officially became written words, and his hatred was publicized to the whole world. In his book, he expresses his anti-Semitic views in detail, describing Jews as a "race" responsible for many of the world's problems.

In 1935, the Nazi government took the next step and passed the Nuremberg Laws, which stripped Jews of their citizenship and many of their civil rights. This was a critical step in the distortion of empathy because once they were no longer citizens, it became easier for the Nazis to view Jews as objects and not as human beings.

Finally, in November 1938, Hilter's hatred manifested into actions and physical violence. That same year, the Nazi regime launched a campaign against Jews in Germany and Austria, knows as "*The Night of Broken Glass.*" Jewish homes, businesses, and synagogues were destroyed, and thousands of Jews were arrested and sent to concentration camps.

You may think this example is extreme, and you will never do anything as horrible as Adolf Hitler. But know the psychological equation that gives birth to the distortion of empathy is the same for you and me.

It starts with thoughts, which eventually have to be expressed in words. Once you can articulate your thoughts in words, it's only a matter of time before they give birth to deeds.

And in our current world, most of those deeds happen at your keyboard.

The distortion of empathy always leads to pain and death - the death of a friendship, death of a relationship, or the death of someone's character. And that is what we now have online. We have a world of online verbal homicides. We stab and kill each other with our words. And every once in a while, it crosses the virtual world into the real world, just like it did for David Molak.

What upsets me most about David's story is when I did my research on him, you could just see he was a good guy. And the first thing the trolls did to make him an avatar was compare him to an ape. I didn't even know him, and it hurt me to read it. It hurt, because most of us can relate to the insecurities we felt when we were young. I am sure he was no different, and his bullies picked something to make fun of that none of us have any control over how we look.

And by doing this, his bullies stripped him of his humanity, and it ultimately gave birth to death.

So, what is the antidote to the distortion of empathy?

There is a famous saying in the business world, which says, *"Your network is your net worth."* It means the more people you know and the more meaningful relationships you have, the more success you will have in business.

This principle is the same in our interpersonal relationships. And in the context of the distortion of empathy, I would change it to,

The more people you know, the less hate can grow.

The way we stop the distortion of empathy is to turn into people who are hungry for other people's perspectives. Parents, if you are raising your kids in a bubble, know they are the most likely candidates for distortion of empathy.

So many people talk about diversity, but few can really explain why you need it. This is not the sole reason, but it is a very important one. Only when you have a large dataset of different people's perspectives do you start to have more empathy for people generally.

For example, I used to be very judgmental when I would see homeless people in the downtown area where I worked. I am embarrassed to admit it, but it's true. However, there are two people who changed my perspective on this topic. It didn't come from a book or a Ted talk. It came from knowing two real people, and although they are totally different, their stories taught me the same important lesson — Knowing someone's story is the birthplace of empathy.

The first person was a man named Michael, a resident at San Antonio's largest homeless facility, Haven for Hope. I was running a nonprofit foundation at the time and Michael had applied for a grant to help him with an invention he created while at the homeless facility. When I heard Michael's story, it made me understand that homelessness is something that can happen to anyone. You take a couple of wrong turns, and life deals you a couple more, and suddenly, you are on the streets. To know him and put a face to the topic of being homeless made my heart softer. We approved his grant, and I am still rooting for him.

The second story is of a relative of mine who was diagnosed as a paranoid schizophrenic. When this person was diagnosed, I started going to a support group specific to this illness. In that support group, I learned an estimated 26% of homeless adults[20] in the US live with a serious mental illness.

This changed me on the inside. Before this support group, my empathy for homeless people was distorted. After this group, when I saw a homeless person walking around, talking to themselves, I no longer saw them as just "a homeless person." I saw the face of my relative.

This is not an easy thing to do, but this is what we must do to eradicate the distortion of empathy. We must become people who seek to find and gather more people's perspectives than just our own. The more different they are from us the better.

In general, I don't believe in silver-bullet solutions. However, of the three Deadly Ds, the Distortion of Empathy has the highest leverage for making the world a better place. The Distortion of Empathy is where the problem begins. The other two Ds enable bad behavior.

If we as a society can fix this distortion, people won't even consider going near the other two Deadly Ds.

Only then can we make the Deadly Ds a thing of the past.

CHAPTER 3

The Last Word & Mob Rules

"Never wrestle with pigs. You both get dirty and the pig likes it."

George Bernard Shaw

In Greek mythology, there is a famous story of Sisyphus, a former king condemned to roll a boulder up a hill for eternity. And every time he got close to the top of the hill, the boulder would roll back down.

This is the perfect illustration of what it is like to engage a troll and to try and get the last word. Just like Sisyphus, you will work yourself up, expend all your energy, only to see your effort roll back down the hill. Allow me to explain.

When I was an account manager, we used an internal system to communicate with our customers. The customers would log onto our portal and open a trouble ticket if they needed help with something. The ticket would then get passed to the appropriate person who could help them. Technical issues would go to engineers, billing issues to the billing team, etc.

What no one told me when I started was that all the remaining trouble tickets in the gray area were sent to us account managers. We also got all the other tickets that were first sent to other departments but for some reason the customer was not happy and would not consider them resolved. It basically made us the cleanup crew.

Let me walk you through an example. I had a customer named Dave who once downloaded a program he had never used before to his server. This program was not from Rackspace. It also caused his website to stop working and caused the server to crash. Dave immediately submitted a trouble ticket and was very upset. The ticket got routed straight to a Rackspace engineer who identified the problem, removed the program causing it, and got the website back up. Our engineer told Dave his issue was resolved, and the site was back up. Dave responded back, telling our engineer he was unhappy that this occurred and wanted a full month of his service refunded. Our engineer then punted the trouble ticket to the Account Manager. That account manager was me.

I politely responded to Dave, telling him that because the problem was caused by something he did, I could not give him a refund. Dave responded back and said this was not true. I responded back and told him nicely that I had verified with the engineer that this was in fact the case. Dave responded that this was unacceptable service and that it was our fault his website went down.

At this point I, had two options. I could respond back and continue this dance, or I could punt Dave to my manager, who would probably do another version of this same dance. A smarter account manager would have probably just passed it off to their manager, knowing full well they were just going to cave and give Dave his refund. But I couldn't do that. I had my pride and it's the principle that matters. So, I took the fool's route and reentered the dance.

So, I responded, and he responded. Then I responded, and he responded. For several days, we went back and forth. By the fifth day, my queue was getting backed up with other customer issues and this guy hadn't lost any steam. That is when it hit me. This guy was not going to stop. I did a quick search on his other issues and found the same pattern of behavior.

That was the day I discovered *the Last Word Principle*. When dealing with trolls, you will NEVER get the last word. Never ever. If you are prepared to respond 1000 times, they are prepared to respond 10,000. If you are prepared to respond for a year straight, they are prepared to respond for a decade. You will never outlast them. Their very lives revolve around this engagement, and they will sacrifice everything in their life just to beat you. Is that the kind of person you want to go 12 rounds with?

The famous Albert Einstein definition of insanity comes to mind, the one that defines insanity as "*doing the same thing over and over and expecting different results.*"

Going back and forth with the troll in your comments section or anywhere else online is insanity. Is that how you want to spend your time? The answer is no, you don't. So just accept you will never get the last word, maintain your sanity, and move on with your life.

BOAT BOY

To really hammer this point home, I want to tell you briefly about another Rackspace customer we had when I was starting out. To this day, I don't remember what his real name was or what his company did. All I remember is that we called him Boat Boy. Boat Boy was a Linux customer who actually lived on a sailboat. True story.

What was also true was that he had spotty internet out there on the high seas, was always breaking his server, and was trying to get us to do stuff we weren't supposed to do for free. He was a classic troll, and because he lived on a boat, he was probably a lonely guy; our Rackspace team was the only human connection he had. This is funny to me because I know someone who spent some time on a boat, and she told me there is always something to do. I guess that includes breaking your website and harassing Rackspace employees.

It didn't matter what we did or said, it was never good enough for Boat Boy. He was out there, waiting at all hours of the night and day for his internet connection to come back so he could respond to us with his new strategy to get us to do free stuff for him.

I tell you this story because I want you to picture Boat Boy the next time you are tempted to respond to someone trolling you in the comments section of your social media. I want you to picture someone who is literally out floating on the ocean with nothing but time. Time to counter your post or your comment.

And once you accept you are dealing with someone with whom you will never get the last word, you will finally break their spell and be set free.

THE MOB RULES

A couple of years ago, I was watching the news with my father and there was story about a Twitter storm that broke out because someone said something that offended everyone. After the story was over, my dad asked me to explain to him what a Twitter storm was.

I gave him the basic explanation that Twitter was a social media website. What made it special was that you could only type a certain number of characters and that was it. And a Twitters storm is when all of a sudden, lots of people are talking about a post or issue. When I was done, he understood but I felt strangely unsatisfied with my explanation.

So, I added, "A *Twitter Storm* also has a dark side. A *Twitter Storm is a rioting mob.*"

"*What do you mean?*"

"*Well, do you remember during the Rodney King riots when we all watched that white truck driver get pulled out of his semi-truck and beat up for no good reason other than he was in the wrong place at the wrong time?*"

"*Yes, I remember.*"

"*Well, that is a Twitter storm. It is the internet version of watching someone get jumped by a mob and not being able to do anything about it. If someone doesn't like you or what you stand for, they can incite an online mob to come after you and hopefully destroy you. The mob behaves online exactly the way they do in real life, only they use words instead of Molotov cocktails to destroy things.*"

"*You don't get to defend yourself or explain your side of the story. You just get dragged out into the digital streets and beat senseless.*"

So much for the public square.

Understanding the Mob Rule is the final piece I want to show you in how trolls think and what influences their behavior. And just like real world rioters burning down buildings, the rules are almost identical.

1.   Trolls love to hunt in packs.

2.   Your facts don't matter.

PACK HUNTERS

You've heard it said before, "*There is safety in numbers.*" Well, online, we can amend this statement to read, "*The more people I convince to attack this person or thing, the more acceptable it must be.*" In the story of David Molak, I saw this show up two different times in the written accounts. The first was in the initial social media post. So many of his fellow classmates joined in as they verbally berated him. Then, in another account, he gets dropped into a text thread of about ten people. He literally got dropped into a conversation he didn't ask to be part of where ten other guys were there waiting to attack him via their cell phones.

How does this happen? How do seemly innocent children feel emboldened to join the crowd and pile on. Some of them will naively say, "*we are just teasing*" or "*it's just a funny prank.*"

What they don't realize is that they have joined the mob. And when the mob sets its sights on something, it normally results in nothing good. They have joined the pack and are hunting *their prey. The scariest part about it is that most of the time,* they know exactly what they are doing. Yet somehow, they cannot seem to stop for fear that the pack will turn on them for not joining in. "*Well,*" they tell themselves, "*At least I'm not the only one doing it.*" And the cycle continues.

YOUR FACTS DON'T MATTER

In the same way you can't reason with a mob in real life, it is also true for the online mob. You would never run up to rioters who are looting a store and setting it on fire and say to them:

"Wait, stop! I have new information. If you knew these new facts, you would change your mind about burning down that building."

To which they respond:

"Lorenzo is really making some excellent points. Maybe we shouldn't set this building on fire but instead sit down and learn more about his point of view?"

Yeah right.

Brick to the face. That is what you would get. The mob doesn't care about your data, facts, or your new information. Always remember these four truths when the mob comes at you:

1. When the mob attacks, your facts don't matter.
2. When the mob attacks, your data doesn't matter.
3. When the mob attacks, your truth doesn't matter.
4. You cannot reason with the mob.

When the mob sets out to destroy something, they are operating in the herd mentality and have surrendered all reason and logic. So, save your breath and walk away from the keyboard. To engage the mob is to make yourself a human pinata.

The best-case study I have seen of attempting to fight the mob is the best-selling author of the *Harry Potter* books, J.K. Rowling.

(**Disclaimer**: I am not writing to weigh in on the issue at the heart of this story but to dissect what happens when an online mob is engaged, so we can draw the lessons out.)

The reason I use this example is that for many years I have followed J.K. Rowling on Twitter and I have always been impressed with how she handles trolls. If you come at J.K. Rowling, just be warned you better bring your A game. If your grammar is bad or you use the wrong "their" instead of "there," she was going to call you out. She is so quick-witted that I used to just read her tweets for pure entertainment. I also love anyone who refuses to be bullied.

Then, in 2020, she retweeted an article in which she disagreed with the author on a gender-related issue. Her statement caused an uproar in the transgender community.

The online community quickly attacked her. Then J.K Rowling did something I would have advised against. She did what she always did: she stood her ground and decided to go toe-to-toe with the mob. A gangster move, no doubt, but one I would advise anyone reading this book to avoid at all costs.

For weeks and weeks, the online community, the press, and celebrities spoke out against her. Then she did something I wasn't expecting. She released a very long article she had written in response to the incident. She talked about very personal issues and even abuse she had experienced, which shaped how she thinks and her opinions. She referenced several scientific studies to show she was not dealing with anecdotes but with empirical data.

The article was extensive, balanced, and thoughtful.

And no one cared.

At least, no one in the online mob, that is. After she released it, it caused even more uproar. Why? Because J.K. Rowling is a brilliant, thoughtful intellectual and she was now dealing with the wrath of the online mob. She tried to fight the mob and the mob always wins.

She was trying to have a civilized dialogue with the mob and the mob does not want discourse, it wants destruction. She tried to reason with them, and they didn't care. Even when she went to Plan Z, which was to spend hours and hours writing a response logically laying out her position, it didn't matter. It was ignored because the mob doesn't care about your past abuse, your scientific studies, or your charitable efforts.

They want your blood and total submission. Nothing less.

So, you may be asking yourself, "*Why do I need to know any of this? I'm not online fighting with people on Twitter or getting attacked for something I said. I will never need to use this information because I normally just mind my own business online.*"

Well, that is exactly what I said and is exactly what I did. I used to mind my own business online too. Then one day, trouble found me when I wasn't looking for it. That trouble was a troll.

And I would soon find myself in a fight I never wanted to be in.

SECTION II

My Fight with a Troll

CHAPTER 4

A Troll in Sheep's Clothing

"War is the unfolding of miscalculations."

Barbara Tuchman

In 2013, I became CEO of Geekdom. For those not from San Antonio, Geekdom is the largest technology-focused coworking space in downtown San Antonio and one of the largest in Texas. For those not familiar with what a coworking space is, just think of it as a gym membership for geeks, tech startups, and entrepreneurs.

Our mission was to create a local tech ecosystem where the next 10,000 technology jobs could be created. This mission was very personal to me, and the main reason I took the job. Back when Rackspace hired me in 2001, it literally changed my life. It was San Antonio's first true tech startup, and when I was hired, I was 20 years old, I had no college degree, I had no tech skills, and had no business acumen. Rackspace took an inner-city Hispanic kid with nothing valued in the business world and it gave me a career, business skills, a network, and so much more.

This is the power of one tech company, so I was excited to become Geekdom's CEO. I took the job because I knew every tech company we helped succeed had the same potential as Rackspace. And when one of those little startups got to the point where they hired their first employee, I knew they could be hiring the next Lorenzo Gomez.

That mission set my soul on fire.

Two or three years into being CEO, things were going amazingly. Business was good, and we were growing like crazy. When I stepped in as CEO, we had about 500 members and occupied two floors of the Weston Centre in downtown San Antonio. Our growth was so explosive our owner bought an eight-story historic building a couple of blocks away called The Rand Building. We moved in and grew even faster than before. We eventually took over three floors of the building and built a brand-new event space at the street level to accommodate the number of events being hosted by our members.

Around this time, we grew our membership to around 1700 paying members, hired about ten or so employees, and had hundreds of companies either officing there or represented by a membership. Our growth was so unexpected, and because of our momentum, the other five floors of the building filled quickly with other tech companies who wanted to be part of the movement.

Everything was up and to the right. Peachy keen, as we might say in Texas.

And that is when I got trolled.

One of our customers was a guy named Johnny. Johnny was a 65-year-old gentleman who had an online store selling used car rims. Specifically, he sold the shiny chrome ones known in that world as "spinners." As a result, he had the nick name of "Johnny Spin." His business did so well, his wife, Tammy, helped him run the operation. Johnny Spin had been on the internet since its creation and, because all his income came from the online world, he was very skilled at using the tools of the internet.

He knew exactly how to use SEO (search engine optimization) to make sure his website was always on page one of Google for his business search terms. He knew how to use social media to build online engagement and bring in sales leads from advertising online.

In short, Johnny Spin was a professional, not only as an entrepreneur, but also as a user of all the tools the online world created. In this sense, I really admired Johnny because it is not easy to make a living online. Not only was he doing it, he had done it for years.

Up to this point, I was really happy to have Johnny in our community. He was friendly and always willing to help make the space fun and welcoming.

All was going well until one fateful day. An incident occurred that changed everything. Johnny had a run-in with another customer, a young man named Scotty. Scotty was 18 years old, new to entrepreneurship and working on a startup that was too advanced for my simple brain. It had something to do with big data and influencers, two things I would roll my eyes at and think, *"There is no future in that."*

From the first time I met Scotty, I really liked him. He was friendly, helpful, and he loved to make people laugh. He was the kind of guy who wasn't afraid to do the dishes in the kitchen if no one had done them and help us clean up the space if we needed help. He also practically lived at Geekdom. He was chasing his dream, and I admired that for someone as young as he was.

Then an incident happened.

A STRAY DOG

Johnny Spin and Scotty were both at a coffee shop not far from Geekdom. There was a stray dog that wandered into the shop. Scotty said jokingly, *"Someone call the dog pound."*

Johnny Spin was not happy about this comment and was very offended, as a dog lover. What Johnny didn't know was that Scotty was also a big friend of animals. The year before, he had lent my ex-wife and me his cat traps so we could trap the stray cats around our house to get them spayed and neutered. He regularly offered this type of help to anyone who needed help catching strays to get them fixed.

But like all great misunderstandings, no one bothered to stop and understand. Scotty told his joke, Johnny Spin told him off, and Scotty fired back. Tempers flared and they almost came to blows. Luckily, no dogs or humans were physically hurt in this story, but as I would soon come to learn, a huge storm was brewing because of this one innocent joke about a dog. So, if any of you think of John Wick going on a murderous rampage to avenge the death of a dog is absurd, I'm here to tell you I have witnessed a real-life dog feud. You can't make this stuff up.

After they got back from lunch, the entire Geekdom staff was talking about the argument. My operations manager, Daniel, came up to me later in the day and told me Johnny Spin had pulled him aside and told him his side of the story.

That is when it escalated; he told us he wanted us to kick Scotty out of the space. I told Daniel to tell him we were not going to kick out Scotty, and I had two reasons:

1. Scotty had been a good, helpful member from the day he joined and had never violated any of our rules.

2. This incident with the dog did not happen inside Geekdom or even on the property where our office was.

As far as I was concerned, this was a dispute between two individuals; they needed to act like mature adults and work it out. Daniel relayed the message, and I crossed my fingers hoping this would be the end of it. I could not have been more wrong.

Johnny Spin got really spun up when we told him no (pun intended). He seemed to turn into another person and started bringing it up constantly like he was trying to build a case for why we should get rid of Scotty. Frankly, it was really starting to annoy me. He reminded me of a child who was not used to being told no. The kind of kid who, when you tell them no, escalates it and throws such a wild fit you eventually give in because you want them to stop screaming inside the grocery store.

But I didn't have any kids, so I didn't care how loud the tantrum got. I was not going to do something just because he didn't like the guy.

The second incident came a few weeks later. Johnny Spin was hosting an event in one of our conference rooms. He had invited some of the other Geekdom members who were also friends with Scotty. They didn't know about the feud and innocently invited Scotty to join them to get food at the event. Scotty didn't know Johnny was putting on the event, so he went along.

As soon as Scotty walked in, Johnny saw him and freaked out. He told him to get out of his event. Being surprised and embarrassed in front of his friends, Scotty fired back something and eventually walked out.

Now, Johnny Spin felt like he had his smoking gun. He came to us and demanded we kick out Scotty. I told him I wasn't going to kick him out. We told him it was a dispute between two people, and they needed to work it out.

GETTING PUNCHED IN THE FACE

The best way to explain the situation I was dealing with is the comparison to a gym. If two members of Gold's Gym had beef outside the gym, the gym would not come in and be the referee. They just need to suck it up and be civil while they work out, or one of them needs to find a new gym. It was that simple to me.

Then came the escalation. We told Johnny Spin we were not going to kick Scotty out and that was the final word. This is when he said something completely new to us. He said,

"Well, he is harassing my wife."

I told him, *"Woah, that is a very serious accusation, and we don't take that lightly."*

He doubled down saying, Scotty was harassing her while she was working in their office at Geekdom. I told him that, of course, we would look into it, but inside I was very skeptical. I thought to myself, *"If someone were harassing my wife, it would have been the first thing I brought up, not the second thing several weeks later."* The accusation was also very vague and had no details to go with it. To me, it felt like he was trying to make his feud our problem by creating a scenario that was both very serious and also happening on the Geekdom premises.

My gut told me Johnny Spin was suffering from an extreme case of Liar-rhea.

I asked Daniel and our Director of Sales, Karen, to look into it. I trusted them because Daniel was a wise older man, had been managing people for decades, and had seen every crazy possible situation you could imagine. Karen was a military veteran, incredibly smart, disciplined, and did not suffer fools. They were the perfect tag team.

I told them I wanted them to interview Johnny Spin, Tammy, and Scotty separately and write down their accounts of what happened. This way, I had two written summaries of each interview, one from a man and one from a woman. I wanted to be able to have a witness for each interview, and I wanted the summaries in writing, just in case we had to reference them later. This would mean I would read six versions of what happened.

They went out and executed the assignment like Navy Seals. When I read the summaries, I learned several things.

First, Scotty was the kind of guy who didn't want to have beef with anyone, so he was trying in his own weird way to rebuild his relationship with them. This meant trying small talk when he saw them in the elevator or the kitchen area.

Second, Johnny Spin was absolute in his hatred for Scotty and was totally against anything that would lead to reconciling with him.

Last, but not least was Tammy. In her interview she said point blank "*Scotty is not harassing me. My husband just really hates him. Just please tell him to not say hi or talk to us at all and we should be fine.*"

And there it was. When I read the interview summaries, I was so angry he would throw out such a serious accusation to get someone kicked out. It is the kind of accusation that could ruin a person's reputation and career, but this guy didn't care. That was the moment I had a terrifying realization:

I was dealing with a professional troll.

I knew what I had to do next, and I also knew that it would cause a shit storm for me and my team, but I had to do what I felt was right. From this point on, I wanted everything I did to be in writing so there was no *"He said; She said."*

I emailed Johnny Spin and kept it short and simple. I told him we had looked into the situation, interviewed everyone involved, and determined it was not a case of harassment but a dispute between two members. And because of this, we would not be kicking out Scotty. I also informed him I considered this to be a closed issue and I would not discuss it again.

I hit "send" and then braced for impact. I knew what was about to happen and I knew it was going to get ugly. The email I sent set off a nuclear explosion inside Johnny Spin's head.

He sent me an email that was full of rage later that day. He told me it was far from being a closed matter, and he was going to tell the entire world we allowed women to be harassed. This was the first time he started using the language of a bully, *"You better do X or I am going to do Y."*

I read the email and thought to myself,

"Do what you gotta do bro, but you are not going to bully me."

He also mentioned in his email he was thinking about con-tacting the police to get a restraining order against Scotty. Again, I didn't respond but I thought to myself, "*Go ahead and do that, that is the exact definition of a dispute between two people.*" But I knew he would never go that route.

Later that day I pulled my team together and gave them a high-level summary of the situation, leaving out the specifics. Then I said to them,

"*Most of you have probably figured out what is about to hap-pen to us, but for those who don't, let me just put it out there. We are about to get trolled hard and I'm not going to lie, it's going to suck.*"

"*My mentor Graham has a great saying that I truly believe in: 'Everyone wants to be a valued member of a winning team, on an inspiring mission.'*"

"*I'm here to tell you straight to your face that for the next 2 or 3 weeks, we are not going to feel like the winning team. For the next 2 or 3 weeks, we are going to feel like the losing team. We are going to get punched in the face, mostly online, and we are just going to take it.*"

"*But here is the good news. We absolutely can get through this, and we will. As long as we don't engage the online mob, we will be fine. After it all blows over, we will go back to being the winning team again. But we have to stick to our guns and not engage online. We must resist at all costs, no matter how bad it gets.*"

I went home that day, and I knew we were about to be the center of a fight, but I didn't know when the first blow would come. Johnny Spin did not disappoint. By the time I woke up the next morning, he had launched an online campaign against us via social media.

I woke up to the sound of my iPhone vibrating. I knew what it was before I even opened my phone. I opened it, saw I had several notifications on Facebook and I had been tagged in a post. I could tell the post was getting lots of traction because every time someone commented, my phone would vibrate, which sent adrenaline shooting to my heart. I opened the post, saw it was a picture of Johnny's wife, Tammy, in front of our Geekdom building. Underneath the picture was a very long post telling their side of the story and expressing how terrible Geekdom was.

It took the strength of ten grown men, but I did not read the post. And to this day, I'm glad I didn't. My immediate thought was:

"She never told him what she told us, and then he bullied her into doing this. He probably even wrote that post for her."

I went to the post's settings, untagged myself from it, and closed the browser. As I walked away, I could feel my heart pounding so hard I thought it would come out of my chest.

I went to work and called another huddle with my team. They told me basically the post was going viral and had been shared several hundred times. I was very clear with my instructions.

1. Do not respond to anything online.

2. If someone has questions, send them to me.

Then I said, *"Try not to read the comments, just go help our members and hopefully it will distract you from this situation."*

It was that simple, but there was nothing simple about it.

REPORTERS SNIFF THE STORY

One of the fallouts I had worried about from the start was the media getting involved. Getting a story like this covered by a good reporter versus a bad one is just a roll of the dice. And since we were such a well-known company downtown, I knew they would all at least look into it. I knew there was a possibility, but I had underestimated Johnny's ability to aggressively sell the story to get coverage. Wrong move on my part. Johnny Spin was a cold-calling machine. He picked up the phone and called every reporter who would bend their ear his way. I feel if he expended that same energy in selling his products, he would have been a millionaire but, oh well.

Then I got a call from one of the local media outlets I knew. They wanted to meet with me. There were two reporters, and they came down to Geekdom to sit down and chat. For this story, we will call them Jack and Jill.

As soon as they called me and told me they wanted to discuss the situation, I sat down and wrote a short simple statement. It read as follows:

"We have looked into the situation and determined it was not harassment but a personal dispute between two members."

That's it. No more, no less.

I had wanted to add "*and we are happy to prove it in court*" but in a situation like this, the more words you use, the more questions you invite. I wasn't trying to prove our innocence via the media. Remember, the mob doesn't care anyway. I just wanted to shut it down from the get-go.

I had worked with most of the reporters in town often because Geekdom and the foundation I ran were constantly launching new programs, so reporters were always writing about what we were doing. But I knew if they asked me for a quote, they were going to be unhappy because I was not going to discuss anything more than the short statement I had written.

Jack and Jill walked in, and were friendly as always. We went into a conference room I had reserved and sat down to chat. I was very nervous because my imagination was going wild. I expected them to start interrogating me good cop/bad cop style. I printed out my one sentence statement and had it folded in my back pocket. My right hand was at my side waiting for them to ask me for a statement and I was ready. I felt like Clint Eastwood, about to have a showdown in a spaghetti Western. Just as he would stand there fondling his revolver, there I was, ready to quick draw my one-sentence statement from my back pocket.

What happened next really surprised me. Before I could even say a word, they just casually started debating each other on if this was really a story or not.

Jill: "*Jack, is this a real story? Angry guy posting all over Facebook - are we really going to cover this? Let the tabloid outlets write about it, not us.*"

Jack: "*I know, but it's getting a lot of attention and lots of people are talking about it.*"

Jill: "*If we cover this, it means we need to start covering every incident that happens at every local company. Are we prepared to start looking into every incident at USAA, Denny's, or Rackspace? Maybe it would be different if he was claiming sexual harassment, but the language is so vague. Lorenzo, are they claiming sexual harassment?*"

Before I could open my mouth, she said, "*Never mind.*" They went back and forth for about ten minutes.

It was actually fascinating for me to sit there and watch them go back and forth. They had such a great rapport. You could tell they really respected each other's opinions and could be intellectually honest with each other. They must have worked together for years. I was just shocked they were letting me witness it.

As they sat there deciding the fate of this story, two details seemed to jump out at me. First, even though the post was allegedly written by Tammy, they kept referring to "*him.*" That told me Johnny was really the one behind this campaign and he was probably doing everything he could to not let anyone talk to Tammy. He wanted to control the story completely. He was such a bully, I thought, even to his wife. I bet he bullied her into doing this whole thing.

The second thing I noticed was Jill's comment about harassment versus sexual harassment. In all the communications Johnny Spin had sent out, he never said sexual harassment. I bet he knew how serious it would make the entire situation and people would want to hear from Tammy directly; that is not what he wanted. However, I realized he would be more than happy if people just assumed it was sexual harassment to keep the story going and stoke the fire. He would be happy to not correct them if it helped his cause. It made me want to throat punch the asshole.

Finally, I tuned back into the debate.

Jill: "*Jack, obviously whatever you decide I am 100% with you. You want to write the story let's write it. I'm just saying that it fails so many tests for me as a journalist.*"

Jack: "*No, you are right. I agree — let's pass. Lorenzo, thank you for your time. If anything changes, I will give you a call.*"

We got up, shook hands and they left. I had my statement locked and loaded, but I never even had to pull it out. I was kind of shocked, but soon I wouldn't be.

Over the next few days, Johnny Spin called and emailed every media outlet he could. News, TV, radio, everyone. But not one of them took the bait. All of them passed.

All of them but one reporter.

CHAPTER 5

One Bad Journalist

"Conflict is inevitable, but combat is optional."

– Max Lucado

I'm not one of those people who thinks all the media is bad. In fact, in my career, I have met more great journalists than bad ones. The problem is the bad ones are really bad and they do a great disservice to the entire industry.

Like many things in life, it's a complicated subject. I like to compare it to good and bad police. Many times, I have heard this interesting quote from police when they talk about bad cops; it goes like this:

"No one hates a bad cop more than a good cop."

It's a great line and you could put it on a Hallmark card, as far as I'm concerned. But there is one problem with it. If I were in a room full of police officers and I asked the question:

"Who here agrees with this statement: No one hates a bad cop more than a good cop?"

I would wager good money every police officer would raise their hand. Then if I asked the question immediately afterward:

"Great. Now, how many of you good cops have ever turned in or reported a bad cop?"

I bet you all $500 in my savings account you would hear the sound of crickets. Why? Because no cop wants to be the guy who turned in another cop. If it was that easy, they wouldn't need a department called Internal Affairs to help regulate them.

Again, personally speaking, in my life, I have actually met more good cops than bad. One of the most traumatic seasons of my life was in middle school. It was a school so bad it made my life a living hell for three years. And thanks to one amazing police officer in the school who took action, peace and order were brought back to the school, and I didn't have to live in fear anymore. But I'm not naïve, and I acknowledge this is not everyone's experience. I know the bad ones are out there, and one bad one can end your life literally.

This is how I view good journalists and bad journalists. It's hard for good journalists to call out a bad one in their profession because the ramifications are so far-reaching, eroding credibility for the entire profession. And unlike the police, journalism has no version of "internal affairs" to keep them honest. You get one bad reporter, people can claim "fake news," and it spreads like wildfire. It's a profession where you really can't afford any bad actors. It's like the airline industry. You cannot afford to have your pilot have one bad day in the cockpit.

The good reporters are seekers of truth, great storytellers and seek to be as objective as possible while telling the public what they need to know, no matter how hard the truth is. The bad ones are power hungry, self-righteous and out to prove something to the world. They believe they are the high priests of the First Amendment and if you question anything they do, you become someone who seeks to obstruct free speech and the truth; therefore, you must be punished.

The bad ones were the English nerds you knew growing up who went on to become journalists. When they get the power to decide what goes into print and what doesn't, they become drunk with power, and their egos take over.

Simply put, the bad ones are bullies. They use articles, columns, and commentary to bully you instead of pushing you into a locker and taking your lunch money.

And as fate would have it, the next bully in this story was a reporter named Rita.

RITA THE REPORTER

The first time I met Rita, she seemed nice enough. She came off a bit shy and soft-spoken but the more I spoke to her, I could tell she was not shy at all. Rita had transferred from somewhere up north, where she covered lots of gang activity, which I thought was really interesting.

A local news outlet hired her to be their new reporter solely dedicated to covering the technology and entrepreneurship beat. Naturally, since I was CEO of the biggest coworking space for technology startups, she called me to schedule a meeting to introduce herself.

We sat down in my office. She pulled out her phone and asked me if I would allow her to record our conversation. I said sure no problem. This was, however, a new experience for me. I never had someone ask me to record a conversation, which definitely changed the tone of the discussion. We talked for an hour about the local San Antonio tech scene. I told her who else she should talk to and what companies I thought were doing great work.

Then, at the end, we started talking about my experience with reporters. I told her this was the first job I ever had where I spoke to reporters regularly and it was a new skill I was having to develop. I told her my experience for the most part was really great and I was just happy to have people covering the companies we were trying to help at Geekdom. Then I said:

"There is one thing I will say I have noticed about some reporters. For the most part, they have been great to work with. But there are some reporters who really give me a bad vibe. It's like they approach every person as if they are secretly an arch-criminal and every company or program is met with mistrust. Almost like they are thinking, 'You are clearly doing something wrong, unethical, or illegal, and it's my job to expose it.' Everything is a scandalous expose, and they are the ones who need to uncover it."

"Those are the only reporters I don't like being around."

"Yeah, I can understand that," she said.

Then, about two months after, we had our first run-in with each other.

A CARDINAL SIN

During this time, I had a team of about ten people working for me. One of those employees was a manager who really wasn't working out. They weren't a fit, and I had taken too long to fix the situation. I needed to be a professional and let them go. Thankfully, before I could do it, the person quit.

That made me happy because I would rather them tell their next employer they decided to leave rather than they were fired. To me, it was a non-issue. These kinds of situations happen every single day at every small business around the world.

Then I got a call from Rita. She wanted to know what happened and what the story was on why they left. I told her,

"There is no story here, it just didn't work out and they moved on."

But as soon as I told her it wasn't a story, I could tell she was offended. Then I saw the kind of reporter she really was. She instantly suspected I was hiding something and began digging into "the story."

There was nothing to be told, so you can imagine my shock when I saw her post a story online about it. The story was about how there was a leadership change at Geekdom, but no one would comment on what happened. She went on to say she went to find the LinkedIn profile of the person who left, and they had not updated what their next job was.

I could feel her implication that this was suspicious and we were hiding something. Then I did something she did not expect. I picked up the phone and I called her editor.

"Gary, this is Lorenzo. I wanted to chat with you about the story Rita just posted about Geekdom. I really think this is not worthy of a story and I'm so confused."

Gary then said:

"I understand but for us, it is news. Think of it this way. It's like the coach of a football team stepping down and there is going to be a new coach. That is all we are writing about is the change."

I said:

"I get that but there is one problem. The way the story is written really casts a shadow on the employee who left. If I were their future employer and I read that article, it would make me not want to hire this person because of how it's written. I want anyone who has worked for me to be able to get their next job with no issues. I want them to be able to walk right back in the doors of Geekdom and get a job with any of the other hundreds of companies here. And this article will prevent them from doing that."

Gary said:

"That's a fair point, I will ask Rita to change the story."

Gary asked Rita to change the story and she did. And guess what? The employee got another job with one of the companies in our coworking space. I felt I had done a good thing in helping remove obstacles for my old employee. The only problem was, it was going to cost me big time.

I was naïve at the time, and I didn't know I had just committed one of the biggest cardinal sins you can commit against a journalist - making them **change their story**. I would soon discover I had created a new enemy for life.

THE ONLY STORY WRITTEN

Back to Johnny Spin and his public campaign against Geekdom. For the first few days, he tried and tried to get anyone he could to write a story about us, but no one would do it. You can imagine Rita's excitement when Johnny Spin called to pitch her the story. By this time, she avoided me and would do anything she could to not talk to me.

It should not have surprised me when my phone vibrated to notify me I had received an email from her about the situation. She had written about ten or so questions she wanted me to answer, and she told me she was doing a story on the harassment allegations. Then, she pulled a sneaky reporter move I had been expecting. She gave me a deadline of a little less than half a business day. This was her way of saying, *"You better respond or I'm writing the story anyway."* It is not uncommon for reporters to have tight deadlines, but this was different. This was a power play.

No problem, I thought. I hit reply to the email. Then I copied and pasted my one-sentence statement:

"We have looked into the situation and determined that it was not harassment but a personal dispute between two members."

I could tell from the questions she sent in her email that this would be a lopsided story with just Johnny Spin's perspective. So, when I sent her my statement, I smiled because I just knew in my gut she was not expecting me to respond that way. My statement ignored all her bullshit questions and signaled I was *not going to play her game.*

Then I just sat and played the waiting game. I thought for sure she would get her story up first thing the next morning, but I underestimated her eagerness to punish me.

A few hours later, the story ran. My phone buzzed as someone forwarded me a link to it. Thankfully, Rita worked for an online newspaper, and you need a subscription to read. And since I was a cheapskate and refused to pay, I couldn't get past the paywall to read the article. It was probably a blessing in disguise.

Unfortunately, one of my teammates did pay for the subscription and pulled up the article on their laptop so I could read it.

As I read the article, I could feel my blood pressure rising. I knew the story would be a hit job but this, to me, was easily the most unethical article I had ever seen.

First, she didn't include my quote at all. She wrote a story and didn't even mention there was another side to it, which I had never seen before. Then, at one point, she completely made up a hypothetical situation regarding Scotty. It said something like:

"No one knows if Scotty may have taken secret pictures or video of Tammy when she wasn't looking."

No evidence, no quotes, just pure speculation. I was so angry, I got lightheaded. I went for a walk to clear my head and regain my composure. At least it's behind the paywall, I thought. Most people won't be able to read that trash. It still hurt, though; I couldn't believe someone could so effortlessly lie and make things up.

Johnny Spin enthusiastically shared the article on his feed and tagged us every chance he got.

The story didn't cause too much damage, but it allowed Johnny Spin to keep the social media outrage flames going.

I thought, *"I have bigger fish to fry, but I will deal with you later, Rita."*

THE BREAKING POINT

During this time, I am proud to say that I never violated the "don't read the feed" principle. But nothing is ever as easy as it seems. Even though I wasn't reading the posts and comments, my friends, family, customers, and, most importantly, my teammates were reading them.

I made a point to talk to anyone on my team who wanted to discuss what was happening but mostly, it was to let them vent. No one wants to be on the losing team and they were starting to get discouraged and demoralized.

I was acting as the world's most unqualified counselor to them and as they vented, they told me bits and pieces of what people were saying online. The same thing happened when I would speak to my friends. They would call to encourage me, and they did. But they would inadvertently say something about a comment or a post someone made about the situation I had not heard yet, and it would linger in my mind. All of this was rattling around in my brain like an emotional wrecking ball.

And as the things people said began to replay over and over in my head, my resolve started to crumble. I started to tell myself:

"You need to do something. You need to fight back."

It felt like the right thing to do. I had to show my team I wasn't afraid to fight or to protect our reputation. But as they say in Alcoholics Anonymous, *"that is stinkin' thinkin'."* I knew intellectually that responding online was the wrong move, but my emotions had kidnapped the driver of my brain and they were now driving the bus. I told myself,

"I have to defend our good name!"

Almost as soon as my resolve collapsed, something happened.

I was miraculously saved.

The moment I decided to respond online, my phone rang: Like an angel from heaven, the voice was that of my friend and mentor, Graham Weston.

He said, *"Lorenzo, I'm so sorry you are going through this situation. I'm calling to check in on you and see how you are holding up."*

"Graham, thank you so much for the call. You have no idea how much I needed it. I haven't done anything stupid yet, but I don't know how much more I can take. I think I am going to respond online and set the record straight."

He fired back instantly.

"Don't do it, Lorenzo. It's a mistake. If you need to take action, do this. Make a list of the small number of people whose opinions you really care about, and then call them on the phone."

It was pure genius. With one tactical suggestion, all my determination and courage came back. I swallowed his advice like Popeye swallows his spinach and I hung up the phone a new man. I knew exactly what to do.

First, I made the list of anyone who was a company partner, mentor, or sponsor. I added another small group of people who weren't necessary in those three categories but who were still very influential in the tech scene and more importantly, people whose relationships I valued.

Then, like all good sales reps, I picked up the phone and started smiling and dialing. The conversations went something like this:

"Hey John, It's Lorenzo from Geekdom. I'm calling you because you are someone I really value and I wanted to tell you about an online attack we are experiencing right now. Before you hear it from the rumor mill, I wanted to tell you straight from the horse's mouth what is happening and answer any questions that you have."

It worked like charm; better than a charm even. People always say the cliché that more communication is better; well, in this instance, it was spot-on. Everyone I talked to was beyond supportive and thankful I had called them so they weren't in the dark wondering what was really going on. I could also tell they were all imagining themselves in my place and they applauded how Geekdom was handling the situation.

There wasn't a single bad phone call. I had managed to do two very important things with this one tactic.

1. I put to rest the emotional feeling I had to *"defend my good name and the good name of the company."* I had defended our good name, but in a constructive, healthy way and to the right people who mattered. Not online.

2. I now enabled a small group of supporters to tell people in their networks the other side of the story and defend our good name for us.

This was the turning point for me and for Geekdom. After this move, I told my team that if they felt like they needed to take action, they should do something similar. Before, we felt like we were tied up and just letting the world punch us in the face. Now, we had been set free to fight back, but in a healthy way outside the online world. It was a huge emotional victory and after this, we all knew and believed we would eventually ride out this storm.

But I was also about to learn a new lesson in this fight, a lesson that no one tells you about. The lesson of success. And if you allow it, victory can poison your heart as much as defeat.

A WEAPON CALLED SILENCE

As I began to think more clearly, I realized just how truly bizarre and odd this entire situation was. For one, the entire time Johnny Spin was alleging we allowed his wife to be harassed, he was still coming into Geekdom. He even paid for his Geekdom membership for another year and a half after that. I don't know about you, but if my wife was harassed somewhere or by someone, you can bet your bottom dollar I would have nothing to do with that person or company.

It also really stuck out to me that somewhere along the way, this entire feud shifted from being Johnny Spin versus Scotty to Johnny Spin versus Geekdom. All his efforts seemed to be directed at us, not the person he was claiming was harassing his wife. It made me realize that because we were more well known in the San Antonio community, he wanted it to be him versus the big bad company. It was all about getting the maximum amount of attention.

I even heard that for the next few years, every time there was a new reporter who moved to town to cover the tech scene, he would meet with them and pitch the story. He didn't pitch them a story about Scotty harassing his wife; he pitched them a story about Geekdom allowing harassment.

Just to give you some perspective, the entire length of time from the moment Johnny Spin launched his campaign against us to it quieting down was maybe three weeks. At the time, it felt like two years, but it really was a relatively short period of time.

I don't know for certain, but I am pretty sure Johnny Spin expected me to fold after the first two days. Given what I would later learn about previous companies he trolled, they all eventually gave in. But I was determined to let him know I wasn't going to play his game or take his bullshit.

So, after the second day had passed from the initial post, he sent me an email that was very amusing to me. The email read as if he had won the battle and was now offering me terms for my surrender. He was saying stuff like, "*I didn't want it to come to this...bla bla bla.*"

I read it, filed it in my Johnny Spin folder in Outlook, then went about my day. The next day, I got another email. I promptly ignored that one too. And as the days went by, he became this strange mixture of angry, frustrated, and desperate. I could hear it in the language he was using. The worst part is, it was starting to make me happy.

I realized more than ever that a bully is so used to getting their way when they don't, it rocks the very foundation of their existence. A child screaming, yelling, and kicking at a store wants you to submit and comply. They are not prepared for you to walk away. It takes away all their power. I know that's probably considered child abuse, but you get my point. I also knew that the more I ignored Johnny Spin, the bigger the tantrum he would throw, but by this time, I didn't care. I knew my plan was working.

Eventually, he got desperate and sent me an email basically saying, "*WHY WONT YOU TALK TO ME!*"

I giggled as I read it. I knew without a doubt I was now psychologically running the chessboard. I had taken a tactical psychological advantage in the situation and, as the great David Goggins would say, I was about to take his soul. I also knew the dramatic temper tantrum would soon follow. And it did.

The next day, he went to social media again. He wrote another long post about how we didn't care and how no one from our company would even talk to him. He had no idea however, that he had already lost the war. No one cared. Not even on his side. Johnny Spin had shot his shot and you only get to cry wolf one time.

If the first post he launched got shared hundreds of times with thousands of likes, this post got maybe a few dozen clicks. His last-ditch effort was to call a good friend of mine and try to get him to broker a meeting between us. I could tell he was desperate to save face and tell people he had come to some resolution. But I was in no mood for reconciliation.

My friend called and told me Johnny Spin wanted him to persuade me to meet with him. He knew I wouldn't, but he wanted to be respectful and relay the message anyway.

And that is when I remembered something I heard online. It was a quote that said:

"The opposite of love is not hate. The opposite of love is indifference."

I realized my silence was hurting him. My silence was the only thing he had not anticipated, and it was the worst blow I could inflict on him. It said to him, *"you are less than nothing to me. You don't exist at all."* It is suffocation by silence. And to a bully, this is emotional manslaughter.

I smiled and thought to myself, *"You thought you could bully me?"*

"Well, choke on that silence you son of a bitch."

Then I responded to my friend with a smile, *"Tell him no thank you."*

About a year later, I heard Tammy divorced him.

CHAPTER 6

Dark Thoughts, Old Feelings

"Do not say, I'll do to them as they have done to me; I'll pay them back for what they did."

-Proverbs 24:29

I wish I could tell you that throughout this ordeal I became the stoic leader who kept his composure, wrestled his emotions into submission, and became the Hispanic version of Ernest Shackleton. But I am not that person, and it is not what happened. It was a very dark couple of weeks for me emotionally.

In middle school, I had gone through a season where the gang violence around me was so bad I walked around a complete nervous wreck. Every day, it was me wondering if today was the day I was going to get jumped by 20 guys and beaten to a pulp. And the fear that comes from not knowing what is going to happen next is something that hangs heavy in your gut. It's a mixture of fear and helplessness.

This is how Johnny Spin's online attack made me feel. Helpless in that you don't have any control over what is happening online and who is going to say what or do next. It's like telling someone you are going to punch them in the face when they least expect it. It's the anticipation that drives you crazy. That was the feeling I had not experienced since I was 12 years old. And every time my phone buzzed, that old feeling would shoot through my body. I had gone to therapy to get that fear under control, and now this campaign against Geekdom was bringing all those feelings back to me; it made me very angry.

Outwardly, I was able to execute the right tactics to help us beat this troll but, on the inside, I was a complete wreck. I had allowed the seeds of hate to take root in my heart and they were starting to grow into full-sized trees.

As the fake story spread online and more people chimed in, I got angrier inside. I started having nightmares and fantasies of hurting Johnny Spin. In one of them, I fantasized I made him bite the curb on the side of the street, then kicked his head into it just like I had seen in an old movie. In another fantasy, I called three people I knew I could trust, and we kidnapped him; wearing ski masks, we beat him until he was unrecognizable. I am ashamed to admit this to you in this book, but it's the truth. In my mind and in my heart, I wanted him to have a permeant scar to remind him daily of our fight.

One morning, I got on my knees to pray, and I started crying in my rage. There is a line in Psalms 3:7 where King David cries out to God saying, "*Arise, LORD! Deliver me, my God! Strike all my enemies on the jaw; break the teeth of the wicked.*"

But I didn't want God to break his teeth, I wanted God to break his life.

The greatest revenge story of all time is *The Count of Monte Cristo* by Alexandre Dumas. I remember reading it when I was about 12 years old, and I was so impressed at how patient Edmond Dantes was in getting his revenge. He waited decades before he crushed his enemies.

There is one vivid scene I remember most from the book. He had just rebuilt a ship for a financially ruined old friend. As the man weeps with joy, not knowing who it was who gave him this life-saving gift, The Count of Monte Cristo is standing far off watching. He makes a declaration that this is the last good deed he will ever do, and now his life would be dedicated to revenge. He stares off and says one of the most gangster lines of all time:

"And now farewell to kindness, humanity, and gratitude. Farewell to all sentiments that gladden the heart. I have sub-stituted myself for Providence in rewarding the good. May the God of Vengeance how yield me His place to punish the wicked."

That is how I felt as this online war came to a close. Sure, I had objectively defeated Johnny Spin and had outlasted the storm. It also gave me great pleasure to know my silence would bother him for a very long time. I should have been happy and grateful nothing worse had happened. I should have been celebrating because the good guys won, for once. But I wasn't happy. I was angry and hate was now brewing in my heart.

I was not done by a long shot. I still had one more score to settle.

And her name rhymed with *Rita*.

WHEN THE BULLIED BECOMES THE BULLY

After the Johnny Spin smear campaign blew over, Rita and I were pretty open about our dislike for each other. I couldn't look at her without getting angry, and she hated that her article didn't take me down. She knew she would still have to work with me, given that she was the tech reporter.

One of my startup friends once told me she was talking about me and said:

"I know Lorenzo doesn't like me. But I used to cover organized crime, and one gang threatened to kill me, so I am not scared of him."

But I had no intention of laying a finger on her. Not a single word, good or bad, would be uttered from my lips in her direction. I was going to use my new weapon of choice — silence. Only this time, I was going to use it with the patience The Count of Monte Cristo had taught me all those years ago. My plan was simple.

Rita failed to realize that, as a reporter covering the tech scene, an unspoken currency exchange existed between our worlds, whether she wanted to admit it or not. The two companies I led constantly created new initiatives and launched new programs. We wanted reporters to write about them, and they were always looking for stories to write about. They had something I needed, and I had stories they needed.

Obviously, you can't make a reporter write about what you are doing, but I knew that with the volume of exciting stuff we were doing, the local tech and business reporters would cover a large percentage of it.

As CEO, it was business as usual, except with one small twist. Whenever we did a press release or launched a new program, I just left Rita off any communications. I would send out press releases to every single reporter except her and, in some cases, for the really big stories, I would choose another news outlet to break the story. This last part is pretty normal.

I knew eventually, her editor and publisher would notice everyone else was writing about the cool things we were doing; everyone, that is, except Rita. Eventually, they would start asking questions, and when they did, it was going to be very awkward for her.

Several months into my strategy she got clever and pulled one over on me. She emailed and told me she wanted to start over and interview me about the state of the tech scene.

At the time, I was overconfident, and I thought: "*It's just one story. I will just be really careful about how I answer; I'll be fine.*"

Rita was more clever than that. She came in for the interview, pulled out her phone just like last time, and asked my permission to record our conversation. I didn't see any issue and agreed as I did before. She had more questions this time, and the interview went longer than I thought it would. Still, there was nothing abnormal about it.

She left, and I thought for a second maybe we were going to move past our feud. But that thought was short-lived. About a week later, someone sent me an article Rita had written about something in the tech scene, which had nothing to do with what we discussed in our interview. She quoted me as if I was commenting on that specific story.

"*Son of a Chupacabra*," I thought. She used things she had recorded me saying in the interview out of context. She pulled this move one or two more times, and it worked beautifully. From the outside, it looked like she was communicating with me and calling me regularly to get quotes. But what no one knew was that she wasn't.

As she was writing these articles about the tech scene, she dipped into the recorded interview we did and pulled a quote from me that she thought would fit the story. I was super pissed off, but I had to hand it to her — it was a pretty brilliant move. But as I once heard someone say,

"*Fool me once, shame on you. Fool me twice, shame on me.*"

I wasn't going to fall for her trick a second time. Several months later, she tried it again, and I politely declined. It just so happened this was around the time Geekdom was launching some really exciting programs, so I stepped up my strategy to exclude her. By this time, we had been playing this game for a while, and I could tell it was starting to get to her.

She was running out of options and getting desperate. Her next move was to try to go over my head to my boss. She emailed our cofounder and asked if she could interview him for a story. The request never even made it to him because I stepped in before anything could get scheduled. I called his executive assistant and told her what I thought. I said:

"Look, you can schedule this interview, but I wouldn't. This reporter is very shady. Let me tell you what she did to me in my last interview that she recorded. The last thing we need is for him to get misquoted out of context for a story he knows nothing about."

What I said was the truth, but I was also boxing Rita in. And it worked. His executive assistant punted Rita back to me, saying that all media requests needed to go through me. Now, she had nowhere to go, so I patiently awaited her next move.

Finally, Rita emailed me and asked if I would meet with her for an interview on the tech scene. This time, I was ready. I politely responded, saying I was willing to help her, but I would not unless certain conditions were met:

1. I would not meet with her in person.
2. If she wanted me to answer questions, she would have to send them in written form, then I would respond with my answers.

She pushed back saying she wasn't allowed to do that. I thought to myself, *"That's bullshit, you literally did that with the Johnny Spin article, but I will play your little game."*

I responded back with a new list of requests:

1. I would be happy to meet in person, but not alone.
2. The interview would have to take place at her company's headquarters.

3. I would not do the interview unless her editor and publisher were both in the room with us.

4. I would not do the interview unless my PR agency was in the room with us.

5. Unlike before, I would not allow her to record our conversation.

Those were my terms; take it or leave it. She agreed, and a date was set.

As the meeting approached, I called my PR agency and gave them a rundown of what was going on and the entire situation. I didn't hold anything back because I genuinely wanted their counsel. I knew I was being vindictive, but I was also playing by the rules. I didn't do anything illegal, and I felt justified being so guarded around someone who acted so unethically from the start.

My PR team told me not to worry; they wouldn't let anything bad happen during the interview. It was a big reassurance and I really needed to hear it.

The day of the interview, I didn't know what to expect. I walked into the company's main office, a ball of nerves. I felt as if I was about to walk up to a chessboard and face an opponent whom Russian chess masters trained.

I was led into a big conference room, and everyone was already there. The editor and publisher were there; they were friendly, shook my hand, and made polite conversation. I could tell from their faces they were both concerned and confused as to why they were there, but they didn't ask me to explain, which I thought was wise of them.

Rita sat at the end of the conference table, and my PR rep was across from her. I walked over slowly, shook her hand, and sat next to my PR rep. Then, the most unexpected thing happened that I had not planned for at all.

Rita started crying.

"I really don't know what happened or how we got to this place. But I'm really sorry, and I want to go back to the way things were."

As she wept, a thousand thoughts raced through my mind. I wanted to say, *"Why don't you explain the article you wrote about Geekdom for Johnny Spin and how you just made shit up? Why don't we talk about how you used quotes from me out of context so you wouldn't have to talk to me. Why don't we talk about how the things you do are the most unethical behavior I have ever seen from a reporter?"*

But I didn't say any of that. Instead, I did what everyone expected me to. I pretended to be gracious and collaborative. I said,

"Hey, don't worry about the past. Let's leave it where it is and start new. All I want is for people to cover our companies' great work at Geekdom. That is all."

But inside, I didn't mean it. When she started crying, my heart smiled a black smile of revenge. I had waited patiently for over a year to get to this point, and now I could witness her defeat face to face. To make victory sweeter, I could see her weep in front of her bosses and a complete stranger. In my heart, I didn't believe anything she said. I told myself:

"You are just acting to try to get sympathy from your bosses. You don't mean any of this. You thought you could bully me?"

"Well, choke on those tears, asshole."

But that was all on the inside. On the outside, I kept my composure. I answered all her questions and smiled. Then I shook her hand and walked out.

About six months later, I heard she left San Antonio and moved to a different state.

SECTION III

The Playbook: How to Beat a Troll

CHAPTER 7

Pain For Me, Gold for You

"Never let a good crisis go to waste."

– Winston Churchill

It wasn't until several weeks after the dust settled that I finally knew we were out of the woods. It was right around that time I found out Johnny Spin had done this before. There were at least two local companies in San Antonio against whom he launched an online attack in retaliation for something he didn't agree with or to force them do something he wanted. Knowing this made me very thankful I wasn't alone, but it also made me hate him even more.

As I sat and thought about it, I realized if I was the third person he had done this to, there was no doubt in my mind he would do this again to someone else. That is how bullies work. They are always looking for their next victim and the next drama in which they can be the center of attention.

One of the reasons I set out to write this book is because I realized thousands of people like Johnny Spin are on the internet. And just like him, they probably have a pattern of attacking people online. Because of this, I wrote this book so you can see what I did well and where I failed. You have to understand that most people will never experience anything of this magnitude in their life, and I am thankful for that. I, however, was not so fortunate.

In my case, I barely made it out on the other side, and not without some serious scar tissue. So, if I can save you and anyone you know the pain and trauma I experienced, I feel like it's my moral obligation.

When this attack came, I found myself trying to do multiple things at the same time. I was trying to process what was happening as quickly as it was happening and escalating. I was troubleshooting and problem-solving in real-time an issue with which no one could help me. I was also trying to manage my emotions, which were all over the place, while at the same time putting on a brave face for my teammates. All of this while also encouraging them that everything would be ok.

When this online assault came, I was unprepared and had no training. All I had were experiences I remembered as a young man at Rackspace.

In hindsight, it was a traumatic experience. But it is an experience that has become the foundation of this book and I would like to pass on to you and those you love.

From this point, I want to give you the playbook I used to defeat Johnny Spin. So much of this playbook is based on and created from that short window of time when my team and I were getting trolled online. My knowledge of trolls came from my time at Rackspace, knowing how the internet works, and how trolls think.

However, the most critical second half of the playbook came from following my gut instincts, combined with some amazing counsel I received from my friends and advisors. I refer to this in my first book, *The Cilantro Diaries*, as my "Personal Board of Directors." This personal board gave me a tactical advantage over Johnny Spin and Rita because I was harnessing the knowledge of several people as I worked through these complex situations. I will explain their counsel in more detail soon.

So, if you or someone you know is getting attacked, you only need to read the following section and make sure you follow it step by step.

WAR & PEACE

First, you need to get your mind straight. You need to decide right now what kind of online user you are going to be. How you use the web when things are going well, when no one is bothering you (i.e. Peace Time) will have a huge impact on how you behave when you are getting targeted or attacked (War Time).

The main reason it matters is because however you decide to use the internet daily will become your number one habit. And when something dramatic happens, and you need to change that habit, it will be like asking a drug addict to quit cold turkey. Fat chance.

For most of us, we sign up for things like Facebook, Twitter, or Instagram because our friends, family, or coworkers are already on those sites. We want to get more connected with them. Sometimes we check out these sites to see what they are all about and explore them.

Deciding how you will interact once you sign up will set the tone for the rest of your digital life. This is your honeymoon period, or as I like to call it, Peace Time. You join Facebook, connect with your family and start sharing birthday pics and asking for plumber recommendations. It's Peace Time. My challenge is for every one of us to adopt a Peace Time playbook that sets us up to use the internet with our best foot forward.

PEACE TIME RULES

Below are some examples of how I decided to use my social media. They are Lorenzo Gomez's Peace Time rules for using social media when I am not getting attacked by trolls. This is by no means an exhaustive list, but it will give you a feel for how I have decided personally to use the internet. I am also in no way judging anyone who uses the internet contrary to my way.

Here are some high-level examples:

1. **Updates** - I use social media to get an update on my friends and family's lives. Example: When they have babies, who just graduated. I am embarrassed to admit it's the primary way I now remember people's birthdays.

2. **Work Content** - I post things related to my books and events where I speak. Also, encouraging wins or progress on my next project.

3. **Tone** - For the most part, I want my pages to be encouraging and uplifting. So, as a rule, I don't post anything that makes fun or belittles anyone.

4. **Passions** - I post mental health-related articles or videos because I have a mental health company. I also love to eat and explore cities, so I post a lot about that.

5. **Hobbies** - I post lots and lots of pictures of my cats.

How I use social media has changed over time. When I was in my 20s and 30s, I used to post all the time. I am in my 40s now, so I try not to post if I can help it.

Here is a list of things I do not post and behaviors I do not engage in:

1. **Politics** – Anything related to politics (exception for some local issues, as I consider most of them to be non-party specific)

2. **Religion** - I am an active member of my church but I do not post anything religious I feel will invite debate online. (Exceptions for things like a good book I would recommend to people)

3. **Useful Apps** - I do not use SnapChat or TikTok. I don't have anything against them, but I tried them out and they have no practical use in my life. Facebook and Instagram are the platforms of my generation; Snapchat and TikTok are the platforms of the next

generation and the generation after will have something new.

4. **Website Boundaries** - I have a Twitter account, but as general rule, I do not have the Twitter app installed on my phone. I only check it on my laptop. This is a personal boundary because when I am on Twitter, I can feel my blood pressure rising as I scroll and realize it is not good for my health. This is mostly because if I go out of my way not to follow someone, I can still end up seeing what they say if someone I do follow retweets them.

5. **Personal Profile Boundaries** - I do not allow negative comments on my feeds or posts. If someone posts something designed to pick a fight with me or someone else, I delete it.

6. **New Tech** - I don't use platforms I don't fully understand yet. If I am interested in something that seems difficult to understand, I will find someone I trust who uses it and ask them to walk me through it before I decide to try it.

Why is this important?

This is important because these rules have now become my daily habits. They are second nature to me. However, if you get attacked or trolled online, the intensity of what happens is multiplied by a thousand. So, my Peace Time rules give me the foundation and support to withstand anything bad or negative that should happen to me.

WAR TIME RULES

This is where the playbook really begins. You are being attacked, harassed, or targeted in some way. You didn't ask for it, or maybe you did. Either way, the fists are flying, and strong words are being thrown your way. It is time to engage the War Time rules, aka, this playbook.

It is my great hope you read this book well before you ever get attacked. However, if you are in the middle of an online attack, it can absolutely help you. It's just twice as hard. It's like trying to learn martial arts in the middle of your first bar fight.

It is better to clearly look at this problem when you are not being harassed or bullied. Then you can see why the plays make sense and how simple they are. And although they are simple concepts, some of them are also very tough to execute. They are tough because they cut straight to the core of our human nature. They poke and trigger deeply emotional parts of our being. Anytime that happens, you are tempted to throw logic out the window. But if you are reading this book and you have arrived at this section, you are more prepared than 99% of the population.

So now let us complete your training. The next section starts the ten principles, aka plays that make up our playbook. They are very simple and straightforward.

I promise you will come out on top if you follow this roadmap. Every time.

CHAPTER 8

The Toughest Row to Hoe

"Where there is no occasion for expressing an opinion, it is best to be silent, for there is nothing more certain than that it is at all times more easy to make enemies than friends."

– George Washington

PLAY 1: ASSESS & DECIDE

I have a great friend named **Alex Baldwin**.[21] He is one of the most brilliant tech guys I know and an even better friend. Back in 2018, President Donald Trump was engaged in an online spat with the actor **Alec Baldwin**. In the middle of their back and forth, Trump tweeted at the actor and misspelled his name, typing Alex instead of Alec. When my friend Alex woke up the next morning, he saw his phone had run out of battery. Apparently, after Trump accidently tweeted at him, he received so many notifications, his phone vibrated so much it drained the battery. He told me he received over 10,000 messages as a result of the mix up. And most of them were not very nice. To put it mildly.

After he was able to get back online and into his phone, he figured out what happened. And because my buddy is an awesome human being, he laughed it off, made a joke about the president's misspelling, and went on with his life.

Here is the point. My friend Alex had to evaluate the situation and decide how he was going to respond. He realized this whole thing was an accident, and he wasn't actually getting trolled by The President of the United States. After he assessed, he made a decision. He decided to poke fun at the situation, instead of jumping into the fight. He also recognized it was an accident and it was better to just move on. And in my opinion, he did exactly the right thing.

This is the first play of our playbook. When you get a heated text, comment, or post you are tagged in, you need to *assess & decide*.

DECODING PLAY 1: ASSESS & DECIDE

Assess the situation to figure out what is actually happening. Then, decide your course of action, how you will respond, or if you will even respond. Are you going to decide to start trading punches with the troll? You must determine if this is a short-term thing or if it's something that you can ignore. If so, I strongly suggest you take a page from my friend Alex Baldwin and walk away. Let it pass.

If it turns out to be a bigger, more long-term attack, you need to buckle up. If you assess and determine you are experiencing a targeted campaign someone or a group has launched against you, your tactics have to change completely.

In short, the rest of this playbook is for you. The remaining nine plays will walk you through what you need to do, and more importantly, why you have to do them.

Follow them closely, because they may save your hide.

PLAY 2: START THE CLOCK

We have all seen this famous scene at the Olympics. It's the 100m hurdles and all the athletes are in crouching positions ready to go. Then, the starting gun goes off and they all shoot out like rockets. In a very similar way, when you get attacked online, there is an invisible, inaudible digital starting gun that goes off and a race starts. Only in this race, you may not have been asked to participate in it. In fact, in most cases, it's a race you get dropped in the middle of whether you like it or not.

That is why this play is very important.

> **DECODING PLAY 2:**
> **START THE CLOCK & NOTE WHEN IT STARTED**

You need to know when the attack started. Somewhere out there, someone posted a comment, a picture, wrote an article, or did something targeted and aimed at you with the intention of taking you down.

When this moment occurs, the clock has started for you. If you go to work, come home, and log onto Twitter only to learn someone posted something about you that morning at 8am, well, that is when the clock started.

You need to note the time the clock started because your goal is to create as much distance between you and the start time as possible. So, take the actionable step, find out when that moment was, and write it down somewhere. Time, Date, Year, and day of the week. You need to note the time as specifically as you can. If I were you, I would write it down on a Post-It note and tape it to the bathroom mirror or your monitor. It could be as simple as this:

Monday, July 8, 2020 8:30am CST

It may seem silly, but I assure you it is not. It is very important to stare long and hard at that time. Why? Because every single second ticking between that moment and the one you're currently in is a very good thing for you.

I will explain why in the next few plays.

PLAY 3: DON'T RESTART THE CLOCK

When you initially get trolled, there is always an inciting moment when it leaves the world of hypothetical threats and becomes a real attack. This is one of the most important milestones to note while going through this ordeal. Why? Because, as we just explained, this is the moment when the clock starts.

So here is the next play.

Once the clock has started, you do not, for any reason, want to restart it. You have now just been thrown into a 12-round boxing match, the likes of which you have never seen. However, unlike a real-world sport, the way you win is to do the complete opposite of what you would do in a real boxing match.

First, for all 12 rounds of this boxing match, you are going to get punched nonstop. You must not, under any circumstances, punch back. Every time you punch back, you *restart the clock*. In this analogy, it means you start the entire boxing match all over again. You made it to round seven, and you couldn't help yourself? You had to throw a jab? Boom. You are back to round one.

The same thing happens in an online attack. The moment you engage, you invite the wrath of the mob and the trolls to attack you all over again, starting from square one.

Second, in a real boxing match, you want to knock the guy out in as few rounds as possible. You don't want to use the entire 12 rounds. In an online attack, the key is to do the opposite - let the clock run out, and let all 12 rounds pass by. In this online boxing match, the clock is actually your friend.

Last but not least, you are the only one who can restart the clock. You may think the trolls control the clock or the media can, but they cannot. To a certain extent, they can affect it, but they can only slow the clock down. The only person who can restart the clock is you.

If you run the play correctly, you will stand there, take your punches to the face for 12 rounds, and run out the clock. Then and only then will it be over.

What kinds of things do you need to avoid doing? Here are some examples:

- Someone posts something about you, and you respond in the comments.
- You make a brand-new post talking about the situation, "*So, you may have heard...*"
- You share a post, a pic, or a screenshot of something that has to do with the situation.
- Someone tweets at you and you tweet back, in the context of them attacking you.
- A reporter or a news outlet posts an article about you, and you repost to your social to "*set the record straight*"
- A famous person or politician tags you in a post or tweets at you negatively and you respond back.

All these are the wrong things to do. More importantly, all of these events restart the clock. What makes this play difficult is the trolls may distract you because they are in all-out-attack mode. They are aggressive and constantly throwing punches and setting traps for you. You must always remember no matter what they do, they do not have the ability to restart the clock. Only you can restart the restart the clock, and the only way the clock can be restarted is if YOU decide to engage.

Now, there are some rare situations when you should engage and restart the clock, but these are few and far between. For example, if the phone rings and a New York Times reporter wants to talk to you, you may have to engage. However, even then, most of the time, you should still pass.

What if it is the very rare time you do need to talk to the media and restart the clock? If you have to talk to the media and restart the clock, we will cover what you need to do in play 7.

For now, let's keep moving.

PLAY 4: DON'T READ THE FEED

In the famous sci-fi show *Star Trek*, there was always a situation when the Starship Enterprise would encounter a hostile alien spaceship. Right before the bad guys would attack, the captain would yell, "*Shields Up*." At this command, someone would push some buttons to make the ship's forcefield activate to protect the ship from what was about to happen.

As the enemy spaceship fired on them, someone would tell them how much power their forcefield or shield had left. They would yell, "*Shields at 80%, now they're at 20%*." Then finally, "*Shields are down*."

This next play is a very strict command.

**DECODING PLAY 4:
DON'T READ THE FEED**

When you are getting attacked online, do not read any of the stuff being said or written about you. The reason this is important is because every time you read the comments, your resolve gets shot and battered just a little bit more. Your resolve to not engage or respond is your shield and force-field. Every comment you read, every response, every post, it's taking you from 100% shield strength to 0%. And when your resolve reaches 0%, you are going to make a mistake.

This is one of the hardest plays in the playbook you will be asked to do. Why? It is hard because it goes against every-thing your body was biologically designed to do.

In the medial temporal lobe of your brain sits a tiny almond-shaped mass called the amygdala. This little guy doesn't sound like much, but you are very fortunate to have it, because the amygdala's job is to do what your homies do for you on the streets, it "watches your back."

It does this because this is the part of your brain where your fight-or-flight response system is located. Part of the job of the amygdala is to look out for danger so it can keep you alive and procreating. It is constantly looking for threats, even sub-consciously. It is trying to process all the information it can to determine if you are in danger and if so, should you fight or take flight.

So, when someone attacks you online, there is suddenly a tidal wave of hate coming your way. When that happens, your amygdala does what it was created to do - it starts to look for threats.

So, in essence, your amygdala starts yelling as loud as it can to the rest of your brain and body,

"Hurry! Go read all those posts. I need to know if there is a threat in there that I should be aware of. There might be something in there I need to know about! Hurry, go read it."

This is why this play is one of the hardest to follow. In order to overcome this situation, you need to ignore the amygdala and not read anything. In essence, you are fighting against something your body was designed to do and wants to do.

You must resist at all costs. You must not read the feed.

When I got trolled by Johnny Spin, the first thing I did was untag myself from the post and I deleted all the apps from my phone. I did this so I wouldn't be tempted to read the feed. You may need to do the same thing. Some of you may need to put your phone away for a while. Some of you may need to have your best friend or spouse hold you accountable and wrestle you to the ground when you start foaming at the mouth and breaking down, wanting to read the feed. But you must not do it.

Resist at all costs. The reason will become clearer when you read play 8.

For now, let's talk about why, if you can execute these first four plays, you are perfectly positioned to take advantage of our next play.

PLAY 5: A RAGING RIVER

A couple of years ago I read a book that put the flow of information online into perspective. The book is called *Abundance*, by Peter Diamandis. In the book, he explains how human beings have exponentially increased the amount of information being created and it's enough to make your brain freeze and your jaw drop. In the book he says:

"From the very beginning of time until the year 2003," says Google Executive Chairman Eric Schmidt, *"humankind created five exabytes of digital information. An exabyte is one billion gigabytes—or a 1 with eighteen zeroes after it. Right now, in the year 2010, the human race is generating five exabytes of information every two days."*[22]

To give you a sense of how much it's increased since then, here is some data that I pulled from a Forbes article published in 2018:

"There are 2.5 quintillion bytes of data created each day…"[23]

"On average, Google now processes more than 40,000 searches every second (3.5 billion searches per day)."

"456, 000 tweets are sent on Twitter every minute."

"1.5 billion people are active on Facebook daily."

Now let me put it in perspective when it comes to being trolled and attacked online. The internet is a raging river of information. This river is running so strongly that when you get attacked, you can actually use it to your advantage. The play is to let this raging river take everything the trolls are saying about you out to sea as new information is created. So here is our play:

DECODING PLAY 5:
GET OUT OF THE WAY AND LET THE RIVER RAGE

The reason people don't use this is because when you are getting trolled, it becomes personal and you start focusing inward. You think to yourself,

"The entire world is going to see this. They are going to be talking about it all day every day."

In reality, you could not be more wrong. They won't be, as you can see from the data above. The entire world is producing so much content - videos, news stories, and even new online attacks - your specific online attack is a drop in the bucket in the grand scheme of things. This massive, constant creation of information every second works to your advantage. It is working to push your story and your incident out to sea.

Now, let me try to make this even more vivid for you. Let's go back to Play 2, where you noted the time and day as below:

Monday, July 8, 2020 8:30am CST

By the time Tuesday, July 9, 2020 at 8:30am CST rolls around, the internet will have created 2.5 quintillion bytes (or 2.5 exabytes) of data. To put this into perspective, if you were to print out all the data created in one day on standard sheets of paper (8.5 x 11 inches), you would need enough paper to cover the entire surface of the United States over 1. 5 times.

In one 24-hour period, that is how much new information has pushed your incident from the starting clock time, out of people's minds.

So, when a troll creates an inciting event and comes at you, they have just tossed their attack into the raging river that is the internet. And if the issue is big enough and crazy enough, they may be able to keep it top of mind for a while, but they cannot keep it forever. Even if you break down and respond online, restarting the clock, all you have done is bring the issue back for a brief time. But it doesn't matter because no one and nothing can withstand the riptide that is the raging river of the internet.

There are billions and billions of users creating new content every second and guaranteeing your story will get pushed out to sea. It pushes anything and everything out to sea. Now and forever.

It's only a matter of time.

PLAY 6: THE SEA OF FORGETFULNESS

Let's return to the height of my trolling attack back at Geekdom. There were several times during those two weeks I nearly cracked and almost broke all the rules in this book. Specifically, there were two times during those two to three weeks when I almost engaged online and restarted the clock.

The first happened right at the beginning. I had put on a brave face so my employees wouldn't worry about the company, if we were going to go out of business, or anything crazy like that. However, after several days of watching them read the comments, I could see it was bothering them. That is when I started to get upset, not for myself, but upset my team was being innocently dragged through the mud.

I did what I was trained to do, I called one of my mentors, who was an early Rackspace executive. I told him how upset I was watching my team suffer. I vented about how hard the team and I had worked to make Geekdom something truly special and it was now going to be destroyed by some troll with a chip on his shoulder. Our reputation would be ruined forever. Then, my mentor asked me a question:

"Lorenzo, do you remember the Rackspace child porn scandal?"

"What?! No, what are you talking about?"

"Well, back in the early days of Rackspace, we had a huge story come out that one of our customers was hosting child pornography on their website. It was sadly the first big press we ever received as a company."

"Was it true?"

"Yes. One day, we received a call from the FBI and they notified us there was someone on our network who was hosting child porn. We were outraged when we found out."

"We wanted to take their servers down immediately but the FBI directed us to not do anything so they could track down the physical location of the owner and arrest him. They didn't want to spook him. They also told us we could not discuss what

was going on with anyone other than the very few people internally who needed to know."

"So, what happened?"

"Well, someone leaked it to the press, and they called us asking us to confirm and give a statement. We were not allowed to, so we told them, 'No comment.' The press had a field day and roasted us. But they didn't know we had been ordered by the FBI to stay quiet so they could catch the guy."

"Did they ever catch the guy?"

"Yes, they did."

"The point is this - if anything is going to take down a company you would think it's a child porn scandal, right?"

"Yeah, totally."

"Well, guess what? Its 15 years later and no one remembers. And if no one remembers that, they are not going to remember some troll calling you names."

He was absolutely right, and I'm thankful for that advice. I went home and slept a little better that night. It was also the moment I realized what the end goal really was for me.

That goal is our next play:

DECODING PLAY 6:
LET EVERYTHING GET TAKEN OUT TO THE SEA OF FORGETFULNESS, NEVER TO BE REMEMBERED AGAIN

In the same way, the internet makes us forget both the good and bad stories online. There are probably some stories that should be remembered and bad people who should not be forgotten, but sadly, that is not how the raging river works. Whether the cause is good or evil, it will eventually push them all out to the sea of forgetfulness, never to be remembered again.

So, if you are worried about your reputation being ruined, don't. I'm not saying it won't be ruined, but I am saying this: The odds are highly in your favor that the world will move on and people will go back to being obsessed with themselves. Your incident will be ancient history. Those are the odds, and I would bet on it any day of the week in Vegas.

But before we move on, I need to further explain what I mean when I say your story will be "forgotten." In a weird way, it will and it won't be forgotten. I call this mysterious truth,

The Data Amnesia Paradox

What you do online will be remembered forever, but it will also be forgotten.

The first half of this statement is about machines, the second half is about humans. You need to know that everything you do online is captured and stored on a computer somewhere. There is what's called "a log" of everything you do. Even deleted posts, comments, pictures, and videos are not really deleted. They are stored on some hard drive somewhere by the social media and tech companies you use. That is just how it works.

What is also true is that even though everything you do is cap-tured and stored somewhere, remind yourself of the equally powerful truth that human beings have a short memory. The computer is a machine designed to store and keep informa-tion forever. The human brain is designed to only remember what it needs to survive and then forget everything else. Both of these things are true at the same time.

As the raging river of data proves, everyone is trying to get you to read their stuff, buy their stuff, consume their stuff, or share their stuff. So, that thing you did 10 years ago you are embarrassed about and are afraid might resurface online? Well, there is a 10% chance it might. There is also a 90% chance no one will ever remember it again. That is just how the inter-net works. Whether you cured cancer or murdered someone, the odds are more likely no one will remember. Even if they do, and someone brings it back up, it will be flash in the pan, because the raging river will take it out to sea eventually. Nothing can resist its power.

So why am I coaching you to believe the internet will forget? The answer is because I need you to realize this truth long before someone tries to troll you. If you don't know this, when the attack comes, you will make irrational decisions because you think everyone is going to remember what is happening to you.

I can't tell you what the winning lottery tickets will be. But I can tell you with supreme confidence that the internet will always forget what happened, eventually.

PLAY 7: SHORT SIMPLE STATEMENTS

In 2023, Prince Harry released a tell-all memoir about the royal family that sold 3.2 million copies in its first week. The drama in the book ranged from how he lost his virginity to physical altercations with this brother.

Needless to say, the media was sent into a frenzy, and it seemed almost daily I would read a news article about something new and crazy that had come out of that book. Each and every time it did, the reporters would excitingly reach out to Buckingham Palace for a response. And each and every time, their response was the same: nothing, no comment, silence. Then the next article could come out and the cycle would repeat itself. This time, the story was juicier and the allegations more dramatic. Media calls the palace, and they get the same sound of a big fat nothing on the other side.

Why am I telling you this? Because the British Monarchy has been in the public relations (PR) business since before America was born. Technically, they have been in the business of managing their PR since the 10th century. They understand these principles better than anyone. And they know one of the most powerful plays you have available is this.

DECODING PLAY 7:
SHORT & SIMPLE OR NOTHING AT ALL

Friends, take a lesson from Buckingham Palace. They know more about how it's done than anyone in the world. If you look at all the articles that came out after Prince Harry's book was published, you will see one consistent theme that shows you how disciplined the crown is. One phrase will show up over and over again, no matter how scandalous the claim or how outraged the people got. Their only response is this beautiful phrase of two words,

"No Comment."

And since the 10th century they have figured out that if they just don't engage, eventually the story will pass by. One publication which seemed to be annoyed with their strategy wrote:

"So far, England's Royal Family has resigned itself to the ignoble role of punching bag, refusing to hit back publicly."[24]

That, my friends, is the answer to the test. That is what you and I must do if we ever find ourselves in the same situation. Don't hit back, become the punching bag. Just remember, if it's good enough for every King and Queen of England, then it's a good enough strategy for you. You are in royal company.

There are, however, outlier situations when you absolutely have to make a statement. Just know that it is your decision, and it should be your last resort, not your first reaction. An example may be your company is getting trolled, you are the marketing department, and the owner of the company tells you you need to respond on behalf of the company. Or maybe you are walking to your favorite coffee shop and a New York Times reporter jumps out from behind a bush and shoves a voice recorder in your face asking you questions.

These situations are rare and, if you are lucky enough to know someone in the PR business, you should consult with them while writing your statement. For the rest of us who are on our own, you need to keep your statement short and simple.

In the case of my troll at Geekdom, the statement I wrote was one sentence:

"We have looked into the situation and determined that it was not harassment but a personal dispute between two members."

That's it. Short and simple; end of discussion. You have just given the world your side of the story, but you haven't given them anything else to get fired up about.

You also need to be aware that you just restarted the clock.

I have good friend in New York who had a business partnership go bad several years ago. Because lots of money was involved, it got really heated. His partner was not only suing him, but making all kinds of crazy accusations. He got wind that his former partner was going to leak a story to the press accusing him of all kinds of illegal activities, which was not true.

He was going to issue a statement his lawyer had prepared, so I asked to see it. They sent it over and it was a mess. It was three paragraphs and closed with a very open-ended statement like *"You will be hearing more from me in the future about this."*

I called him and said:

"Dude, we needed to perform surgery on your statement. It's way too long, and all it's going to do is invite more reporters to write about it."

"*Also, we absolutely need to cut your last sentence saying they will hear more in the future from you. After you issue this statement, they will never hear from you again because you will consider it a closed matter. All you are doing is inviting any interested reporter to follow up with you in a couple of months.*"

We changed it to say something like this.

"*These accusations are totally fabricated and look forward to proving it in a court of law.*"

That's it. Short and simple. And when the article hit, it was a collective nothing burger, just as I expected. Also, as an extra bonus, it was the same week Jeff Bezos announced that he was stepping down as CEO of Amazon. The raging river was helping us already.

If you stick to these plays, you will ensure the clock runs out and don't invite more trolls to jump on the bandwagon to make things worse. It will also set you up to be in the best position to tackle our next play. And you will need all the help you can get.

PLAY 8: DEFEND YOUR NAME, LOSE THE GAME

I am going to tell you out of the gate that this principle is one of the hardest plays we will discuss in this book. If any of these principles is going to break you, chances are it's this one. How do I know that? I know because it almost broke me.

Why is this principle so hard? It is hard because the desire to defend your good name is so strong when attacked that our natural reaction is to fight back. Under normal circumstances, that would be ok. However, in the online world of trolling it is exactly what you cannot do. Why? Because defending your good name requires you to engage online and violates our play of not restarting the clock.

For me, it went down like this. When Geekdom got trolled, I started off very strong. I didn't engage in any way. I didn't read any of the nonsense people were saying online. Even when the reporters called, it didn't faze me. I gave them my one sentence statement and moved on with my day.

I started to weaken when I saw my team suffering, but even then, I was determined to get through it for them. What broke me was my inner and outer circle. My inner circle of friends and family was unwavering in their support for me. They called me constantly to encourage me and reminded me they were there for me. But in doing this, they would innocently slip up and tell me little snippets of the gory details, something someone had said online or in person who was talking trash about us and the situation.

My outer circle of friends and colleagues did something similar, and it was not with any bad intentions. They would call me to give me information and, in doing so, they were weakening my resolve. They didn't mean to; they thought they were doing me a favor by telling me what they heard and what the word on the street was. Truth is, I didn't need to know and didn't want to know. I just didn't have strong enough boundaries to tell them what I needed, which was for them to stop telling anything. So, I sat there, smiled, and just took it.

Finally, the breaking point came, and I couldn't take it anymore. I was having constant fantasies about posting my rebuttal and lighting everyone and their grandma up in flames. My self-talk got really bad, and it was constantly whispering bad advice to me, advice like:

"You need to fight back. You need to go to war. You look so weak for not fighting back. Who is going to defend you if you don't defend yourself?"

As the voice rang in my head on repeat, I got weak. That is when my resolve cracked. And as recounted in section 2, that is when Graham called and gave me two gifts. The first gift was the reminder that it was a mistake to defend myself.

That is, it was a mistake to defend myself... *online.*

Defending your name is a landmine; if you step on it, it will explode and create so many more problems for you. It will invite much more commentary and opinions you don't want. More importantly, defending your name takes you out of the objective realm of reason, and places you in a state of mind completely and 100% based on emotions.

And my friends, the last place you want to be is in an emotional mindset when it comes to online attacks. Only bad things will happen when your emotions are driving the bus.

This is what Graham reminded me of when he called me. This was his first gift. His second gift was the tactic that gave me the emotional relief I needed. It was the gift that took all my pent-up energy and put it to constructive use.

PLAY 9: MAKE A LIST, THEN ENLIST

When Graham finally called me, I was an emotional wreck. I had allowed my fear and anxiety to run amok, and my brain would not shut off. As I tried to problem-solve what to do, my mental processor kept coming to the same conclusion: Defend your good name. Respond.

One of the reasons I love Graham is that he is a hope dealer of the greatest kind. He is also someone who is always in search of a good business tactic. He has constantly told me:

"Most people spend way too much time on the strategy. What we need to find are tactics that bring the strategy to life."

So, when he called and told me not to respond, he knew I needed something more. He knew I needed a tactic, and that tactic is what I give to you now.

DECODING PLAY 9:
MAKE A LIST, THEN ENLIST

Make a list. Write down the names of all the people whose opinions you really care about. When the list is complete, call every one of them on the phone and tell them what is going on. If you can meet them in person, even better but it's not mandatory. What matters is that they hear it from the horse's mouth. Now that you have talked to them, they have been unknowingly enlisted to help you fight the good fight. This is important for three reasons.

First, it shows you are proactive and not hiding in a corner. Especially if some of these people are customers or business partners, they may be wondering what is going on and some of them may be even starting to believe what is being said about you online. As my therapist always says, *"you can't control what other people do or say, but you can control what you say."* This play allows you to at least tell your side of the story.

Second, as you talk to people, you will start building a small army of advocates for your side of the story. After I called all the people on my list, I started hearing countless stories like this:

"Lorenzo, I heard someone talking about the Geekdom situation and they had it all wrong. So, I told them they had their facts wrong and set them straight."

Nothing is more powerful than having other people defend you and put out the right narrative. Your friends and family actually want to help you, but they don't know how. When you tell them the real story and what is going on, you arm them with a way to help, which they are eager to do anyway.

Third, all the stress, anxiety, and emotional burden of not defending your name will start to go away. It will feel amazing, like a pressure value releasing all the tension built up in your body and mind. As you call and talk to people, you will feel better because you are finally doing what your body and mind want, which is to defend yourself. But you will be doing it the right way. You will be defending yourself the constructive way, which is not engaging online.

The real brilliance of this play is that it gives you real action to redirect your energy. The desire to defend your name online is so great, very few people can escape its gravitational pull. So, when you feel like you are about to break, remember sometimes the old ways are better. This is one of them. But do it face-to-face or on the phone.

Defending your name is a tactic of the physical world. You should avoid any and all forms of it online. This includes posts, comments, replies to things people are saying, and I would even say emails and DMs. All of it.

Once you type one word and hit send, you have reset the clock and have to start all over again.

PLAY 10: TROLLING IS TRAUMA

I read a story a while back that broke my heart. It was about a high school girl who had sent her boyfriend some risqué pictures of herself. Then they broke up and in an act of sheer evil, he posted them online. When the girl saw her pictures were spreading across the internet, she killed herself. What's more, her boyfriend was arrested on child porn charges because his ex-girlfriend was under the age of 18 and he had posted nude pictures of her online.

Everyone lost in this story. If you think this example is extreme, I'm here to tell you it's not. Why would a young girl go to the extreme of killing herself? Why did I fantasize about physically hurting Johnny Spin? Why did I decide to get even and take my revenge on Rita?

We avoid talking about the answer in polite company because it makes us sound weird. We try to ignore it and pretend it doesn't exist, but we need to call it by its real name: Trauma.

That is the hard truth of the matter - Trolling is trauma.

And this leads us to our last play.

DECODING PLAY 10:
DON'T SUPPRESS THE DISTRESS. SPEAK UP & ADDRESS

Plays 1 through 9 are designed to help you while you are in the middle of getting attacked and beat up online. If you keep getting attacked, all you have to do is rinse and repeat plays 1 through 9.

This last play, however, is the most important one of them all. The reason it's important is because we, as a society, do not like to talk about our feelings. We think it makes us look soft or weak. We also tell ourselves this lie: "*Even with the people I know love me, I don't want to bother them with my problems.*"

This lie is killing us emotionally. This last play is the most important because if you have been attacked online, you need to know you have experienced a form of trauma. Not only that, you are walking around with an emotional infection so dangerous that if you don't treat it, it will spread to other areas of your life and to the people around you.

If you or someone you know has been the target of an online attack, you have experienced both emotional and psychological trauma. And because this is a form of trauma our society has not yet figured out how to deal with, it is very important you seek help.

Here is what you need to know about the path forward.

Admit the distress – Admitting you have experienced something painful and traumatic is the start. When you don't admit it, you are walking around in a constant state of emotional isolation. You feel like it is only happening to you. This means you are walking around telling yourself, *"No one understands what I'm going through,"* or *"no one else has experienced what I am experiencing."*

This negative self-talk is bad because, if you say it long enough, it starts to morph into even more extreme talk like, *"If I were gone all this would go away"* or *"these people hurt me, so I should hurt them back."*

In the example of the girl who killed herself, she told herself she was all alone and the only way to stop the pain was to end her life. In my example, I told myself the only way to get relief from my pain was to get revenge and even the score.

Right now, as you read this book, there is some young person alone in their room reading horrible things people are saying about them on their smartphone. They feel like they are utterly alone in the world. But they are not alone.

Emotional isolation is the road leading to self-harm and harm to others, which is why bringing it out in the open is key to processing it. That is step one. In fact, I feel so strongly about this topic I cofounded a company whose sole mission is to "*End Emotional Isolation*." The company is called **WeTree;** it's an online tool to get emotional support from friends and loved ones. If you want to look into it further, you can check it out at www.getwetree.com.

Speak up – You need to tell someone you trust what you have experienced. For those of you reading this book who have been targeted and trolled online, I strongly recommend you seek professional help from a therapist or a counselor. For some, it will be a pastor, priest, or rabbi. For some, it will be talking to your parents or the authorities. For some of you, it means you need to speak up when you see trolling or online bullying happening. Maybe it means notifying a parent when you hear about it. Maybe it means notifying a school. But you need to say something if you see it.

Address – If you don't do something or "address" all of these unprocessed emotions, they eventually come out. Look at my example - I did the classic caveman move and tried to go "*tough guy*" in my situation. I buried it way down and thought I could move on or ignore it by working hard. In the end, it showed up in stress, nightmares, and constant fantasies of anger and rage. I even started grinding my teeth so much I had to get a mouthguard.

The point is, your unprocessed emotions will find a way out. That is how our body is designed. If you don't do it in a healthy way, your body will do it in whatever way it can. Most of the time, that means an unhealthy way. It can lead to classic coping mechanisms like alcohol, drugs, sex, overeating, and all kinds of other activities you engage in to numb your emotions and block out what happened. Don't do it. Get help instead. Learn from me. It took me almost half a decade before I spoke to my therapist about my trolling event. I would give anything to go back in time and process it with him the week it went down.

I am not being dramatic when I say you need to get help. The stories of people killing themselves or others because of this trolling illness is growing every month and year. So, believe me when I say that getting help will save lives.

And one of those lives could be your own.

CHAPTER 9

Boundaries Beat Bullies

"Those who guard their lips preserve their lives, but those who speak rashly will come to ruin."

– Proverbs 13:3

My second book is a mental health book about my middle school, which is located in the most socioeconomically segregated zip code in San Antonio. The title of the book is, *Tafolla Toro: Three Years of Fear*. Tafolla is the name of the school. About two years after I published the book, a local school system called. CAST Schools asked me if they could adapt the book into a play because the book was resonating with their students. I happily agreed and posted something on social media about the upcoming performance. That is when I received a long comment on Instagram from someone trolling me.

It was a guy under a fake name telling me it was gross how I was trashing the school, (I wasn't), and how, instead of being part of the problem, I should go down there and help the kids (which was the whole point of my book, to help kids struggling with fear and anxiety). It also said I was exploiting the school to make tons of money for my books. If I remember correctly, my royalty check that month was for $79. Thanks, Amazon. The post was long and very bitter. Classic trolling.

What did I do? I deleted the post because I don't allow negative comments on my personal profiles. Then, I direct messaged him and said:

"I saw your post. I'm sorry you feel that way, I'd love to buy you coffee, lunch or dinner so I can hear your story and you can hear mine."

To my surprise, the guy agreed. Turns out, he was a teacher at Tafolla. When we met for tacos, my first question was,

"Have you read the book?"

His answer was, "No."

I can't lie. I was a little annoyed that someone would write an angry comment about a book they actually didn't read but I was also trying to keep an open mind.

I heard his story, and I told him mine. The irony to me was that he would have liked the book based on his story. I did learn something else. He was trying to help those kids the best way he could. For him, that meant going back to the school and teaching them. I explained I was also trying to help them but in a different way. I am not a teacher; I am a writer, which is the tool I used. He was angry that I used the school's name, and he assumed it meant I was trashing it. I explained I was very intentional in picking the title of the book. As a matter of fact, most publishers would have suggested I change the name of the book because "*Tafolla Toro*" is difficult to pronounce. But I didn't write the book for mass marketing. I wanted any kid who grew up in that poor Hispanic neighborhood to read it and say, "*Someone out there understands what I am going through.*"

I wish I could tell you that we became best buddies, but that didn't happen. When he commented on my post, he was just a random user, devoid of humanity to me. By sharing a meal, I could avoid the distortion of empathy in my mind. It was my way of bringing the humanity back for both of us. Our dinner was civil, but I could tell I was not going to change his mind, and he was not going to change mine. We had dinner and left. I also invited him to the play so he could see it was designed to help kids. He showed up, but I knew he was never going to be a fan.

The point of this story is this.

When he trolled my post, I had to access the situation and make a decision. I decided I was going to practice what I am trying to preach in this book. My first step was to draw some boundaries.

My first boundary was not allowing angry or hateful talk on my personal accounts, so I deleted his comment. My second boundary was that I reached out to him and told him why I did it, that if he wanted to discuss the situation, which he was clearly upset about, we could do it in person, face-to-face. In today's world, I don't recommend most people offer to meet someone face-to-face; if you do, I would recommend taking someone with you. This second boundary was my way of saying, "*I am willing to engage in constructive dialogue, but I don't engage online.*"

These two simple boundaries completely changed the nature of this interaction. What could have turned into a heated, back-and-forth exchange turned into a mutual understanding. Sure, he is still not a fan, but at least I got to explain myself. Even though I know he still hates my book, after those boundaries, I am able to sleep like a baby.

So, here is my question to you. What boundaries do you need in your online world?

Setting boundaries means establishing guidelines and limits around your online activities and interactions to protect yourself. This includes limiting the personal information you share online, restricting who can access your online accounts, and watching your online behavior.

We need to implement more boundaries than we currently do online. Here are three areas where you should take an inventory and bring more boundaries into your world. This is not a comprehensive breakdown but a starting place for you; these are just three areas I want to touch briefly.

They are the areas of... *people, places, and things.*

DIGITAL PEOPLE, PLACES, & THINGS

THE PEOPLE

We need to start our boundaries journey by talking about *people*. There are probably some people you need to cut from your online world. Some people are doing nothing but brining hate and negativity to the platforms you are using. You need to let them go. For some, it means you unfollow them, but you keep them as contacts; this is most common in Facebook. You need to unfriend some people because, when you really think about it, you are not friends and have nothing in common. In the worst circumstances, there are people you should block.

Some of you may need to make your profiles private so the only people you interact with are the ones you allow in. You have chosen these people to have a window into your life. I have been guilty of not doing this and making my profiles completely open.

If I were starting to use the internet today as a new user, I would probably have everything set to private. Because I am a public figure, I choose to open my profiles to the public, but that means I need other thoughtful boundaries to protect myself. You should have some, too. Here are three questions to consider:

1. Have you ever had to remove any negative or hateful contacts from your social media accounts?

2. Have you ever talked to your loved ones about removing online friends who are causing them distress?

3. Have you ever adjusted your social media privacy settings to limit interactions to those you trust?

DIGITAL PLACES

Next is *digital places*. The principle applies to anyone and all ages. Digital places are the websites, forums, and apps we use regularly. The internet is so big it is a place where you can find anything and any interests. No matter what the interest, no matter how niche it is, there is more than likely an online community just for that specific group.

For example, are you really into 1967 Ford Mustangs? Not the 1966 or the 1968, just the 1967 model. Well, the good news is that an online group probably exists just for that.

This is a good thing because it allows you to find your people and join a community no matter how obscure your interests are. The bad part is, no matter how obscure you or your family's interests are, there are also groups for them. The question is, do you know what those are? Do you know what your family is doing in them? I am not implying they are bad but I am here to tell you that most people are participating in some type of group on the web.

In my research for this book, I compiled a small list of popular websites for families. This is not the definitive list, but it does pose an important question to anyone reading this book. How many of these websites do you even know and do you understand how people use them?

1. **YouTube** - YouTube is a video-sharing platform allowing users to watch, upload, and share videos. It is popular because it offers a wide range of content, including music videos, gaming, educational videos, and more.

2. **TikTok** - TikTok is a social media app allowing users to create and share short videos. It has gained immense popularity especially for its viral dance challenges, lip-sync videos, and comedic skits.

3. **Snapchat** - Snapchat is a messaging app letting users send photos and videos that disappear after being viewed. It is popular because of its fun filters and the ability to communicate with friends in a private, visual way.

4. **Instagram** - Instagram is a social media platform that lets users share photos and videos. It is popular for its visual nature, and it is often used to show-case one's life and interests.

5. **Reddit** - Reddit is a social news aggregation and discussion website. It is popular because of its diverse communities and the ability to share and discuss a wide range of topics.

6. **WhatsApp** - WhatsApp is a messaging app allowing users to send text messages, voice messages, and make calls. It is popular because of its ease of use and the ability to communicate with friends and family around the world.

The digital places we go online have become the new TV of this generation and it is the place where people now invest most of their time. It's important to know what those sites are all about. More importantly, you should have healthy boundaries in place. Here are three questions to consider:

1. How do you balance your time spent on digital places with other activities and responsibilities in your life?

2. Do you know what digital places your loved ones visit, and have you discussed the importance of digital boundaries with them?

3. Are you aware of the privacy and security settings available on the digital places you visit, and have you taken steps to protect your personal information and ensure healthy online boundaries?

DIGITAL THINGS

Digital things are all the ways people access the world wide web. These are the devices people use. You need to know this list is growing every day. It is not just the family computer everyone shares. There are many more ways to access the web and you need to be aware of them. More importantly, you should be asking yourself, asking if you need boundaries around any of them.

In the popular Amazon Prime show, *Jack Ryan*, there is a season where they are chasing a terrorist who is plotting mass terror attacks in the US. The primary way the terrorist communicates is through an online game where you can chat with other players. And although the story is fiction, this functionality is absolutely real and usable today.

I have put together a small list of the most common devices people use to access the internet. Please keep in mind this list is not exhaustive and it varies by region. You will be surprised to learn how easy it is now for people to get online.

1. **Mobile devices**: This includes smartphones and tablets. Mobile devices have become the primary means of accessing the internet globally. According to Statista, in 2021, 53.3% of web traffic worldwide was generated through mobile devices.

2. **Desktop computers**: Desktop computers remain popular devices for accessing the internet, especially for tasks requiring a larger screen, such as work or gaming.

3. **Laptops**: Laptops offer the convenience of being portable, making them a popular choice for students, professionals, and anyone who needs to work on-the-go.

4. **Smart TVs**: Smart TVs have become increasingly popular, as they allow users to access streaming services and internet browsing directly from their televisions.

5. **Game consoles**: Modern game consoles, such as the PlayStation and Xbox, also offer internet browsing and streaming services, making them popular devices for accessing the internet.

6. **Wearable devices**: Wearable devices, such as smartwatches and fitness trackers also have internet connectivity and can be used to access certain web-based services.

I don't point to any of these things to scare anyone. I point them out because knowledge is power and the road to having strong, healthy boundaries starts with knowledge. You need to decide right now for yourself and for your family how you will use the internet and what boundaries you need. You should decide how you access the internet, what sites you will go to, who will you interact with and how. Here are four questions to consider:

1. How do you balance the convenience and accessibility of digital devices with the potential risks associated with internet usage?

2. Have you set clear boundaries for yourself and your family members?

3. Have you discussed with your family or loved ones the importance of unplugging from digital devices for a certain amount of time each day to promote better sleep and reduce screen time?

4. Do you use trusted and reputable sources, such as official app stores, to download and install new apps on your digital devices to minimize the risk of malware or other security threats?

Once you can answer these questions, you will have the foundation for healthy and strong boundaries. However, beware that the internet changes constantly and all these places and things are moving targets. What is popular today will not be tomorrow. You need to stay informed and have real conversations every year about people, places and things in the online world.

CONCLUSION

A New Social Contract

"The secret of change is to focus all your energy, not on fighting the old, but on building the new."

- Socrates

When I worked at Rackspace early on, they had us all take a personality assessment to help focus on the areas in which we were naturally gifted.

The assessment showed I was high in *Positivity*. This means I am generally very encouraging and have an optimistic outlook on life. It's totally accurate and this strength is one of my motivations to write this book.

I am not naïve in thinking we can turn the internet into the magical land of Narnia. I do very strongly believe, however, we can change how people interact online.

We live in the age of the one-liner comebacks. One-liner comebacks are only good when it comes to branding, marketing, and video clips going viral. What are they not good for? They are not good for building relationships or community.

Having a superior argument or better jab has never ever, in the history of humankind, made someone change their position. It only pushes people to their corner to dig their trench deeper. One of my favorite theologians is a man named Dr. Timothy Keller. He wrote a book called *Forgive: Why Should I and How can I?*, and he said something that really resonated with me. He said:

"Win the person, not the argument."[25]

A while back, I was having coffee with a close friend, and we were catching up on local events. At the time, the San Antonio city council was in a contentious negotiation with the police union. I said something really dumb about unions being corrupt and it really upset my friend. The conversation immediately took a negative turn, and he began to explain how much the unions had helped his family up north where he grew up. He was really upset, and I could see the situation spiraling out of control. This friend is like a brother to me and the last thing I ever would want is to hurt or upset him. So, I stopped him mid-sentence and I said:

"I choose you and our friendship over this issue. Honestly, I am sorry I said anything in the first place, I don't know enough about unions to really have a thoughtful opinion and I certainly don't care enough about the issue to have it affect our friendship. I'm sorry for what I said, please forgive me."

In that moment, I was intent on winning the person, not the argument. And that is what we need to bring back to our conversations online.

It is my belief the internet does not have to be a dark, hate-filled cage fight. We can be civil, ask questions, and respectfully disagree. I know we won't change everyone, but we can change our own behavior. This is how movements start - small acts from determined individuals who have made up their mind to be different and act differently.

Cat memes and civil conversations, I would even say passionate conversations, is what awaits us. That's the internet we can have back. But we need to take it back, one kind word at a time.

THE CODE OF CONDUCT

Did you know that during the civil rights movement, Dr. Martin Luther King Jr. and other civil rights leaders developed a code of conduct for people who wanted to participate? During that time, if you wanted to join the movement and participate in protests, sit-ins, and other nonviolent actions, you were required to sign and adhere to the code of conduct.

How incredible is that? Here are a couple of principles Dr. King wrote himself:[26]

- Use active non-violent resistance to evil.
- Never seek to defeat or humiliate your opponent, but to win his friendship and understanding.
- The non-violent resister seeks to defeat the forces of evil, not the person who happens to be doing evil.
- Avoid external physical violence but also internal violence of spirit (hating the opponent).
- Accept suffering without retaliation.

- Recognize that the center of non-violence is the love of God operating in the human heart.

I think these are brilliant examples of what we need. So why reinvent the wheel? The world needs to take a page from this nonviolent, direct-action movement, and adopt its own code of conduct for the online world.

So, I am going to propose my own version of an internet code of conduct. Take or leave it; it doesn't matter to me. Make your own or a better version of what I have presented and that would make me even happier. Why? Because, at the end of the day, all I care about is that someone else out there is committed to changing how they live their online life. Here is my attempt at my code of conduct.

THE 12 LAWS OF ONLINE HUMANITY

1. **Active Empathy** - I will remember at all times there is a human being on the other side of this screen who is worthy of respect.

2. **Empathy Expansion** – The more people I know, the less hate can grow.

3. **Radical Transparency** - I will post as if the people I hold most dear are watching me.

4. **Acknowledge Nuance** - I will never attempt to have a complex debate online, only in person and with people with whom I am in a community.

5. **Disagreement Discernment** - I know I cannot change someone's mind through arguing online. I choose to win the person, not the argument.

6. **Bold Boundaries** - I will walk away, log off, or block anything that steals my peace or gives me anxiety.

7. **Self-Control** - I will not attack even if I am being attacked online. I will guard my words so I do not cause others pain.

8. **Active Encouragement** - I will encourage those around me, those I don't truly understand, and especially those who have been attacked online.

9. **Protect & Assist** - I will speak out against online attacks and abuse by sharing these principles, tools, and plays, with those that are being attacked.

10. **Community Building** - I will use the tools of the internet to build community, not cause division.

11. **Reject Retaliation** – I will not retaliate against any person or persons even if they attack me.

12. **Mob Mentality** – I will not fall prey to groupthink and will not join other people who are attacking someone online. I will not participate in group aggression, even if it's a "joke."

My plea is for you take the parts of my story that worked and use them, and that you take the parts of my story where I failed and don't repeat them. Only then will my pain have been worth something and not in vain.

Finally, don't do it for me or even for you. Do it for the people around you. Do it for the children in your life who need you to be an example of how to act online. Do it for your mothers and fathers who, at one point, were cautious of the internet and are now slowly engaging more and more. Do it for them, and lead by example so we can show the people we love what it means to "*love thy neighbor*" in the real world and the online world.

It is my great hope these principles and stories will save you. I hope they will save someone's reputation. I hope it saves someone's self-esteem. I hope it will save someone's family and marriage. I hope it will save someone's job.

Lastly, I hope that one day, stories like David Molak will become a thing of the past because these principles will save someone's life.

That is my hope. I hope it is yours too.

THE ANSWER — THE GREAT COLLISION

This book starts with a young woman's question: "*What is the biggest difference between going to school now and when you were going to school?*"

I can finally say many years later I have a better answer to that young woman's question, so, I will close this book by taking my do-over.

The biggest difference is the collision of the human experience and the online world. Everyday, these two forces become more intertwined. Your mobile phone has now become an extension of your hand. The phrase "*Google it*" is so common, I think it would give Coca Cola a run for its money when it comes to world brand recognition.

This difference is so great that if you described it to someone 50 years ago, they would have laughed and called it science fiction. Now, with the dawn of artificial intelligence tools, this mixture of human experience and technology is only going to accelerate.

But all hope is not lost. I remember hearing internet titan and pioneer, Marc Andreessen, once say on *The Tim Ferris Show*: [27]

"There is something deep seeded in human psychology where we are always going to invent the thing that's going to kill us"

He continues:

"And then, it turns out it's a tool. It's a technology. It's something that helps us do things in a better way. It's overwhelmingly to the benefit of mankind. And then, we wonder why everybody got so worked up over it."

From Prometheus getting fire from the gods to the invention of electricity, the railroad, social media, and now A.I., all these technologies are tools. They all also have one thing in common - Human beings are the ones who wield them and decide how to use them. So here is the question back to that young woman and the question I now pose to you:

How will you respond to the collision of the online world and the human experience.? It's coming to you faster than you think.

In my personal journey, I was living my life and enjoying my career when the new tool of social media came and declared war on me.

I was bullied and then I bullied back. I won and still lost. In being bullied, I found myself wrestling with anger, bitterness, betrayal, and fear. In becoming a bully, I found myself fighting the deeper emotions of revenge and forgiveness.

For me, I finally had to make the choice to stop playing the movie of what happened to me over and over in my head. I decided to turn it off and move on with my life. This book is a big part of that healing process. I also decided to forgive the people who wronged me, not for their sake but for mine. If I had not, I would be sentencing myself to life in the prison of bitterness. I choose to forgive and be free.

I also choose to forgive myself for the pain I inflicted on my offenders. Without this self-forgiveness, I would be dragging around guilt and shame for a long time. It is my choice to forgive myself and the choice to make better decisions daily to unshackle me from this emotional pain.

So, the next time your worlds collide, what path will you choose to take?

Of course, it is only natural to feel angry and want to hurt back. And when you arrive at this pivotal moment in your life, I hope you will stop and remember something. I hope you will remember David Molak. I hope you will remember my story and the crazy roller coaster ride of emotions it took me on. Let these two stories stand between you and making some bad decisions. Let these stories be a sign pointing you to a better path. The choice is yours to make.

Love is also choice, and it requires zero followers, zero likes, and zero shares. So, make the choice today and turn off the bully in your pocket.

That is how we take back the internet. One person at a time. One person deciding to delete that post or not post at all. One person deciding to encourage instead of criticize, forgive rather than holding a grudge. One person deciding to walk away from the keyboard. One person deciding to log off.

So, this is Lorenzo Gomez III, former troll.

Logging off.

THE BULLY IN YOUR POCKET

The bully in your pocket
Is the monster that you grow,
Spitting hate and fire
For all the world to know.

A dragon who encircles
Every victim that he maims,
A face that's so disfigured,
By a tongue of crimson flames.

He's the demon sent from Hades,
His purpose crystal clear,
To kill the human spirit,
And draw destruction near.

But the world no more can bear it,
Our resistance now decreed,
We set out to hunt and kill him,
Our hearts to be set free.

So, I take this vow today,
To fight with all my might,
To starve the dragon out,
For justice is in sight.

This new decree I pledge to me,
To turn my back on hate,
Abstaining from the endless rage,
A new world order to create.

I choose the path of peace,
Of lavish praise and love,
Forgiveness over vengeance,
And glory from above.

I choose to lift up the stranger,
Whose face remains unseen,
And hope my words have given life,
On both sides of our screen.

By Lorenzo Gomez III

ACKNOWLEDGEMENTS

This book took me about two years longer than I expected to finish. I want to first start by thanking my parents and my family for their overwhelming encouragement and support. Mom, Pops, Danny, Tara, Denise, Marco, Martha, Hector, Sonia, Jeff, Mari, Roland, and Patty. Thank you to my nieces and nephews Andrea, Richard, Devina, Brandon, Markie, Joshua, Noah, Elisha, David, and Oliva. It doesn't matter what the situation is, I know that if I need you to show up it's as good as done. I love you.

Thank you to my personal board members who are constantly there to guide me, give me counsel and feedback. Danny Gomez Jr., Dax Moreno, Doug Robins, Luke Owen, Cody Lockwood, Emily Bowe, Khaled Saffouri, Pravesh Mistry, Jamie Hong, Bill Schley, Jake and DJ Gracia, Steve and Marisa Cunningham, Jeanne Russell and Mike Villarreal.

Thank you to my beta readers whose feedback was so valuable to me and to helping me finish the book. Thank you, Barbara Boyd, Dax Moreno, Bill Schley, David Heard, Doug Robins, Jamie Hong, Jeanne Russell, Shannon Forester Smykay, Steve Cunningham, Sophie and Khaled Saffouri, Sal Webber, David Robinson Jr., Janel Galindo, Danny Gomez Jr and my mother.

Thank you to my Thursday night men's group. Being in community with you has given me new strength at a time when I really needed it. Thank you, Hector Gomez, Cody Lockwood, Jamie Hong, Joshua Collins, Scott Turner, Marcos Hernandez, Steve Cunningham, Thomas Lopez, Brett Dyer, Chad Curtis, Abel Pacheco, Sam De La Rosa, Chris Toland, Ronnie Lott, Salvador Jimenez and Edwin Stephens.

Thank you to all my close friends that are always a never ending well of encouragement and joy. Thank you, Ajay Rayasam, Tommy McNish, Jaime Cooke, Pat Matthews, Joey Boatright, Vlad Mata, Luis Martinez, Alex Baldwin, Carlos Maestas, Jennifer Maestas, Sean Wen, Andrew Ho, Ryan Hunter, Leah Lindsay, Falcon Rubio, Zach Schoenfelt, and Anne Wolfe-Andersen.

Thank you to the Burnt Nopal team for making yet another amazing cover that I am absolutely in love with. Cruz Ortiz and Olivia Ortiz, thank you so much.

Thank you to my longtime friend and teammate Laurie Leiker. I am so happy to get to work with you again and I could not have finished this book without you. Thank you to Ed Rister, Cara Collins, Merritt Weeks, and Jody Hall for being so encouraging to my writing.

Last, but not least, thank you to Graham Weston for the endless support you gave me during this season of my life.

ABOUT THE AUTHOR

Lorenzo Gomez III is a community builder, author, and public speaker who draws from his personal and professional experiences to inspire others. A former director at Rackspace Technology, Gomez is a key player in the development of San Antonio's downtown entrepreneurial ecosystem. He has spent ten years dedicated to building institutions and organizations that have supported the creation of a contemporary community of technology companies. Gomez is the co-founder of WeTree, a new mental health engagement application and co-founder of Confluence Capital Group which invests in community centered real estate.

During Gomez's four years as CEO and four years as chairman of Geekdom, San Antonio's premier collaborative startup community, Geekdom companies raised more than $422M in funding and its startups have created 2,489 jobs.

A trusted advisor to Graham Weston, co-founder of Rackspace Technology, Gomez helped create Weston's 80|20 Foundation, which invests in San Antonio's future by issuing grants to public charities that attract, grow, and retain San Antonio's future workforce and job-creating entrepreneurs. After co-founding the 80|20 Foundation with Weston, he served as its executive director for six years, during which time the organization founded Students + Startups, San

Antonio Startup Week, and helped establish the UTSA School of Data Science downtown. He and other local tech leaders also founded Tech Bloc, an advocacy group for San Antonio's tech economy.

Gomez's most recent venture, Confluence Capital Group, is dedicated to catalyzing and enhancing micro-districts near San Antonio's urban core. The company envisions creating vibrant, walkable neighborhoods and communities that attract young professionals seeking an authentic urban experience, thereby helping San Antonio compete for global talent.

Gomez has authored three books: *Cilantro Diaries*, *Tafolla Toro*, and *The Rack We Built*. All three are Amazon best sellers and the latter two are winners of the 2021 International Latino Book Awards. *Cilantro Diaries* tells the story of Gomez's rise from the stockroom of a grocery store to the boardrooms of two private companies, without a college degree. *Tafolla Toro* tells the story of his turbulent, traumatic, and often violent middle school years in one of San Antonio's most crime-riddled neighborhoods. He reveals the fear, anxiety, and hopelessness he felt as a teenager and how those forces shaped his life until he began taking steps as an adult to improve his mental health. *The Rack We Built* examines Rackspace's growth from a scrappy startup to a billion-dollar technology company and the evolution of its legendary corporate culture.

Lorenzo's driving passion is his love for his hometown, San Antonio, Texas. His goal is to recreate for every San Antonian the opportunities the startup world has provided him in this city on the rise.

NOTES

1   My San Antonio article, "Alamo Heights High School Student Was A Victim." https:www.mysanantonio.com/news/local/article/Alamo-Heights-High-School-Student-Was-A-Victim-of-6743320.ph

2   My San Antonio article, "Alamo Heights High School Student Was A Victim." https:www.mysanantonio.com/news/local/article/Alamo-Heights-High-School-Student-Was-A-Victim-of-6743320.ph

3   https://www.mysanantonio.com/news/local/article/Exclusive-Fellow-students-threatened-violence-6746095.php?cmpid=artem#photo-9223112

4   https://people.com/human-interest/david-molak-family-speaks-out-after-suicide/

5   https://www.mysanantonio.com/news/local/article/Alamo-Heights-High-School-student-was-a-victim-of-6743320.php

6   https://www.davidslegacy.org/

7   Start-up Nation: The Story of Israel's Economic Miracle, by Dan Senor and Saul Singer, page 26

8   Start-up Nation: The Story of Israel's Economic Miracle, by Dan Senor and Saul Singer, page 26-27

9   Start-up Nation: The Story of Israel's Economic Miracle, by Dan Senor and Saul Singer, page 25

10   Start-up Nation: The Story of Israel's Economic Miracle, by Dan Senor and Saul Singer, page 25

11   Facebook post by Cliff Molak https://www.facebook.com/cliff.molak/posts/10205516164573749

12   BBC News article, "Twitter Hack: 130 Accounts Targeted In Attack" https://www.bbc.com/news/technology-53445090

13   United States District Court for the Northern District of California Criminal Complaint, United States of America v. Joseph James O'Connor, Case No. 3:21-mj-70812 MAG https://krebsonsecurity. com/wp-content/uploads/2021/07/O-Connor-complaint.pdf

14   BBC News article: "Briton Pleads Guilty In US To 2020 Twitter Hack," https://www.bbc.com/news/technology-65540901

15   BBC News article: "Briton Pleads Guilty In US To 2020 Twitter Hack," https://www.bbc.com/news/technology-65540901

16   David's Legacy Foundation article: "David's Law One-Pager," https:// www.davidslegacy.org/wp-content/uploads/2018/08/Davids-Law-One-Pager-R2.pdf

17   Cyberbullying.org Bullying Policy https://cyberbullying.org/bully-ing-policy

18   The Science of Evil: On Empathy and The Origins of Cruelty, by Simon Baron-Cohen, page 7

19   The Science of Evil: On Empathy and The Origins of Cruelty, by Simon Baron-Cohen, page 7

20   City of San Antonio Substance Abuse and Mental Health Services Administration article, "Current Statistics on the Prevalence and Characteristics of People Experiencing Homelessness in the United States," https://www.samhsa.gov/sites/default/files/programs_campaigns/homelessness_programs_resources/hrc-factsheet-current-statistics-prevalence-characteristics-homelessness.pdf

21   People Magazine article, "A Man Named Alex Baldwin Is Having A Terrible Day on Twitter Thanks to Trump's Misspelling," https://people.com/politics/alex-baldwin-donald-trump-alec-baldwin-tweet/

22   "Abundance," by Peter H. Diamandis, https://books.apple.com/us/book/abundance/id454135978

23   Forbes Magazine article, "How Much Data Do We Create Every Day? The Mind-Blowing Stats Everyone Should Read," https://www.forbes.com/sites/bernardmarr/2018/05/21/how-much-data-do-we-create-every-day-the-mind-blowing-stats-everyone-should-read/?sh=79ebd-fa60ba9

24   The Hill article, "Royal PR Battle Rages As Harry Spares No One,"

https://thehill.com/opinion/technology/3806617-royal-pr-battle-rages-as-harry-spares-no-one/

25 Forgive: Why Should I and How Can I? By Timothy Keller, page 187

26 Wisconsin Historical Society, The Student Voice vol 1, no 1, June 1960, page 5 https://content.wisconsinhistory.org/digital/collection/p15932coll2/id/50061

27 The Tim Ferriss Show, Transcript of interview with Marc Andreessen https://tim.blog/2018/01/01/the-tim-ferriss-show-transcripts-marc-andreessen/